BUCKINGHAMSHIRE
COUNTY
LIBRARY

CUBLINGTON

A BLUEPRINT FOR RESISTANCE

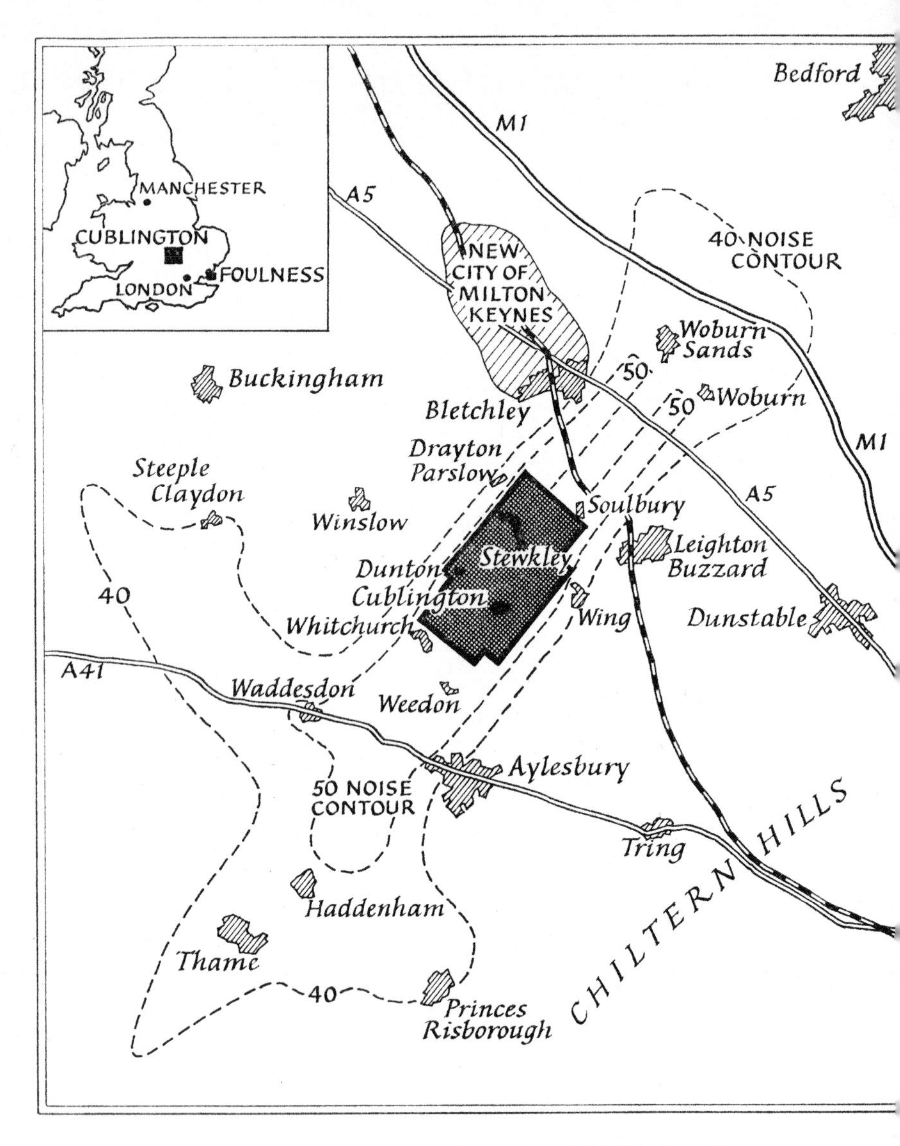

The proposed Cublington site for the Third London Airport, showing the level of aircraft noise expected by the year 2000.

CUBLINGTON

A Blueprint for Resistance

DAVID PERMAN

THE BODLEY HEAD
LONDON SYDNEY
TORONTO

ISBN 0 370 10235 5
Printed and bound in Great Britain for
The Bodley Head Ltd
9 Bow Street, London WC2E 7AL
by Richard Clay (The Chaucer Press) Ltd
Bungay, Suffolk
Set in Monotype Bembo
First published 1973

CONTENTS

ACKNOWLEDGMENTS

Although some Cublington people have paid me the compliment of telling me: 'It's difficult to remember that you were not actually here,' the fact is that I approached the events related in this book as a stranger to them and at a time when the airport battle was over. This meant that I had to rely heavily upon the memories, the advice and the judgments of people who had been both present and active in the anti-airport campaign. They are too numerous to thank individually by name and some of them have said they prefer not to be named. To them all, however, I owe an immense debt of gratitude both for the wealth of information they provided and for their patience and generous hospitality.

Among them are some whom I must name because they went to the additional (and considerable) trouble of reading the many drafts to the manuscript. Those who were subjected to this burden were Desmond and Susan Fennell, Jeremy and Isobel Smith-Cresswell, Geoffrey and Barbara Ginn, David and Helen Stubbs, Bill Manning, Arthur Macarthy, Edward Ashcroft and Dr Jeremy Richardson of Keele University. Geoffrey Holmes and Stan Newens commented on individual chapters of the manuscript and Michael Turner of Sheffield University gave me valuable insights into parliamentary enclosures in Buckinghamshire. They were all unfailingly kind in pointing out errors but, of course, all the errors that remain and all the opinions expressed are the responsibility of the author alone.

I must thank also Mr E. J. Davis and his staff at the Buckingham County Record Office, the staff of the County Reference Library at Aylesbury, the WARA secretary, John Pargeter, and his staff, and Mrs Tiepz, formerly librarian to the Roskill Commission. Last but not least I must thank my two typists, Deborah Hyams and Jane Lush, without whose hard work and good humour this book could not have appeared.

D.P.

Harlow,

December 1972

CUBLINGTON

A BLUEPRINT FOR RESISTANCE

I

A STORY FOR OUR TIME

Cublington was the name of two campaigns. One was a grim affair, an attempted rape of a section of English rural life that was repulsed with the stubborn and inventive resistance that the English have traditionally shown to foreign invaders. The other was happier and far less parochial. It was the remarkable way in which the people of Cublington captured the attention of the world's newspapers and television and, thereby, transformed their local anxieties into a debate about the environment at large.

Cublington itself is a small, straggling village in Buckinghamshire. It lies some 40 miles to the north-west of London in the gently hilly country on the edge of the Vale of Aylesbury, country that has been aptly called 'unmitigated England'. Within its parish boundary is a disused RAF transport airfield, which during its active war years was known by the name of another adjacent village, Wing. The two names have often confused people. 'Cublington (Wing)', to give it its official title, was proposed in 1969 as a site for the third London airport. Eighteen months later it became the firm recommendation of the Commission on the Third London Airport, headed by a High Court Judge, Mr Justice Roskill (now Lord Justice Roskill). It was the proposals of the Roskill Commission which provoked the resistance which is the main subject of this book. It was a resistance which spread far beyond the parish boundaries of Cublington and Wing, for the third London airport at Cublington would have borne little resemblance to its wartime cousin. It would have been a monstrous development suited to the twenty-first century, planned on a scale which probably few people even now have grasped. It would have covered an area the size of Central London within the ring of mainline railway stations. As it grew to maturity, it would have swallowed three whole villages—Cublington, Stewkley and Dunton—and possibly two others—Soulbury and Whitchurch—though not in fact Wing. It would have made 1,700 people homeless and made life intolerable for some 10,000 others.

Its footprint of noise, fumes, fuel dumps, access roads and other impedimenta would have blighted 350 square miles of three English counties. To house its servants, it would have begotten a new city of 275,000 people. Someone in Stewkley, in fact, described it as a dragon in search of a home.

The people of North Buckinghamshire reacted to this threat with a militancy of which ordinary Englishmen are seldom thought capable. In a nation where the farmers are comfortable and conformist, where the tractor-led demonstrations of European workers of the land are regarded with incomprehension, the Cublington protesters were literally outlandish. They rolled their tractors and combine harvesters on to the roads, they lit bonfires and burned effigies. They lobbied the authorities at every level but they also took their case direct to the nation, inviting their countrymen to come and see for themselves what Cublington was fighting to preserve. And the people came—in their thousands. The protesters employed every expert they could afford and when the experts had done, they became their own experts. They were, as it was often recalled, of Buckinghamshire stock, in the tradition of the village Hampdens withstanding the little tyrant of their fields. This modern tyrant, though, was hardly a little one.

To those not directly concerned with this struggle, the other campaign—the assault by Cublington on the news media—must have seemed very much like collaboration. The newspapers, radio and television revelled in this new-found militancy. Reporters and producers assigned to cover the goings-on in Buckinghamshire often acquired an attachment to the area and to its protesters. Some of them came back in their own time to renew acquaintances and lend support. Millions of column inches of newsprint and miles of television film were devoted to this unfamiliar phenomenon of rural protest. Press coverage spilled over from the newspapers into journals like *Punch*, *Private Eye*, *Reader's Digest*, *Women's Weekly*, *Country Life*, *Garden News*, *Amateur Photographer*, *Farmer's Weekly*, *Architects' Journal*, *Building*, *Campaign* (advertising) and *Ringing World* (campanology). The television networks of West Germany, Holland, Canada and Japan showed films made in the villages of North Bucks. The *New York Times* said in an editorial that Cublington's fight was a fight for mankind as a whole. It was world news and greatly entertaining.

The sheer fun associated with Cublington was one of its most memorable features. The cartoonists had a field day. Many of their works were later auctioned by the anti-airport association and raised

£1,850 for campaign funds. The humorists had their day too. 'Cublington' became synonymous for the joke-tellers with a re-enactment of the Peasants' Revolt. Here is what Alan Coren made of it in a *Punch* piece called 'Extracts from the Encyclopaedia of New English Folklore':

'*Cublington Fair.* A regular event, held in Buckinghamshire, and traditionally associated with Airport Bill, the local name for the Devil. (There are analogues at Stansted and Foulness.)

'The Fair begins with a march of Freeholders, usually middle-aged, who parade with quaint hand-lettered signs nailed to long staves which they wave above their heads, shrieking imprecations the while in the hope of averting the curse of Airport Bill, who, they traditionally believe, will appear to them in the form of terrible noises in the sky, and break their greenhouses.'[1]

Yet it was, of course, a serious matter and not only for those threatened by the airport. At its base, the Roskill Commission was an inquiry into the most important planning and national investment decision that Britain will make in the next decade, possibly in the remainder of this century. The treatment of Mr Justice Roskill and his fellow-commissioners in the press was often a travesty of the truth. This was partly a result of the Cublington campaign. The people of Cublington were fighting for nothing less than survival and they waged all-out war on the Roskill arguments that threatened their communities with extinction. It may not have been fair, but it was wholly understandable and it was successful. People still live and farm and worship in the villages of Cublington, Stewkley, Dunton, Soulbury and Whitchurch and they no longer have the shadow of the airport hanging over them.

Cublington was a remarkable study in how to capture the attention of the media, yet it too suffered more than a little in this respect. Although its relations with the press were on the whole cordial, there were also detractors and critics who came down from Fleet Street. Some accused the village Hampdens of being tyrants in their own right. They were said to be trampling down the rights of less vocal communities, especially of that at Foulness, the remote site on the Essex coast at which the third London airport is now to be built. Cublington people were accused of denying freedom of speech to their own minority who wanted an airport. They were accused of being not the mass protest movement they claimed to be, but a small, 'almost disturbingly professional' pressure group masquerading as popular protest. A few newspaper reporters came down to Buckinghamshire determined to expose this masquerade. They attended demonstrations in search of 'real countrymen' (presumably men in smocks with straw sticking out of their ears). Having failed to find any, they talked about it being a purely middle-class affair, the work of the 'gin-and-tonic set'. They enumerated the local landowners who had political influence in London. A television programme on the pros and cons of an airport at Cublington devoted most of its footage to a discussion in the drawing-room at Woburn between His Grace the Duke of Bedford (pro) and Her Grace the Duchess (con). They were all saying the same thing—it was the ex-urbanites, the landowners, the aristocrats and professional people who opposed the airport, not the ordinary people. Regrettably, these denigrators, although a minority of the journalists who wrote and broadcast about Cublington, have had their effect: in many quarters the name of Cublington now stands for middle-class protest. If this book can change that erroneous impression, it will have achieved something.

There were other denigrators who, consciously or unconsciously, ascribed the success of the airport resistance to something else entirely. Two Fleet Street journalists summed it up in these terms:

'It was a campaign based on emotions but carefully controlled and directed by a clever team of professional public relations men working for a high-powered American-based company called Burson-Marsteller. For a fee of £3,500, the PROs chatted up selected Fleet Street journalists, the television companies, Members of Parliament, to create the "right image" of Cublington: the area became an irreplaceable slice of England's green and pleasant land where an airport, any airport, would be a rape of the countryside. The professionalism was, of course, backed up with well organised demonstrations by the airport resistance association's 65,000 members.'[2]

Of course! The 65,000 members must have helped a lot; perhaps they came from Rentacrowd. The remarkable thing about the airport resistance campaign at Cublington is that it did attract 65,000 supporters, that it did raise £57,000—only a tenth of which was spent on publicity or public relations of any kind—and that it did capture the imaginations of a nation and of the world's best newspapers for months on end. It is a phenomenon which was only partly explained at the time when these events took place. That phenomenon needs analysis as much as the story itself needs now to be told as it really happened.

Any competent journalist ought to be able to capture something of the flavour of the Cublington affair and its impact on the nation, but its importance to the people involved, to those who lived and breathed resistance to the airport for more than two years—often to the detriment of their health, their jobs and their families—all of this places a quite different responsibility on a writer. This is mainly their story.

On a spring afternoon in 1971, Cublington won its reprieve. On April 26, the Secretary of State for Trade and Industry, Mr John Davies, told the House of Commons that the third London airport would be built, not at Cublington as the Roskill Commission had recommended, but on the reclaimed Maplin Sands off the coast of Foulness Island in Essex. Within days, mention of Cublington began to disappear from the newspaper headlines. Naturally it was not so soon to be forgotten by the people who for two years had fought for just this decision. But for the rest of us, who had only watched and read and perhaps sympathised with their fight, it soon passed into the limbo of yesterday's news, where fact and argument and one conservation issue and another become indistinguishably mixed. Why then should the wider public remember Cublington?

The first reason is that it was an environmental issue of great importance. This was partly a question of timing. The Roskill Commission made its recommendation to the Government at the end of 1970 which had been designated 'European Conservation Year'. In Britain there was growing—some would say, rather modish—concern for all aspects of environmental conservation and the new Conservative government had lately created a new ministerial overlord with the title of Secretary of State for the Environment. The Cublington story came to its climax at a time when a major test of the Government's intentions and of the nation's concern was needed. But there was more to it than that. The countryside in this quarter of North Buckinghamshire may not be dramatically lovely—though it is near to the justly famous Vale of Aylesbury and the western escarpment of the Chilterns—but it does have much to recommend it to the hearts of the nation. Sir John Betjeman, in evidence to the Roskill Commission on behalf of the Wing Airport Resistance Association, put it thus:

'Just try driving, as I did, with the wind behind you and a western sun on the leaves of elms and oaks, along the pleasant undulations from Winslow via Swanbourne, Drayton Parslow and Stewkley to Soulbury and you will see what I mean. This is what one means by England. To lose this bit of unselfconscious, intimate and friendly country so near to London is to lose part of ourselves.'

It was Sir John who applied to this countryside the phrase (from Henry James) 'unmitigated England'. His was a poet's view of Cublington. But many people who thought in more prosaic terms were saying something similar. The message of the Wing Airport Resistance Association (WARA), the main anti-airport group in the area, was that the loss of this tract of open country, with its relatively well-preserved villages and ancient churches, would be a loss not only to the people living there but to the nation as a whole. WARA's symbol was a set of concentric circles radiating out from a centre obliterated by a jet plane. Their case was that the implications of the third London airport plan rippled outwards from Cublington to a much wider area of national concern. It was a representative battle for the English countryside. It was a battle to conserve an important stretch of country on the very doorstep of the sprawling metropolis, to prevent the creation of a corridor of concrete from London to Birmingham. It made little sense, in any case, to spoil such country in a nation as densely populated as Britain and as lacking in landscape unspoilt by 200 years of industrialisation and urbanisation, when there was the ready alternative of an offshore site at Foulness, which would actually increase the amount of land available.

These WARA arguments were transformed into an eloquent testament of environmental faith by Professor (now Sir) Colin Buchanan, the one member of the Roskill Commission who opposed the choice of Cublington. In his note of dissent to the Roskill Report, he said that an airport at Cublington would be 'an environmental disaster'. Drawing upon his unrivalled experience as a planner, Professor Buchanan—who was at the time Professor of Transport at Imperial College of Science and Technology, London, and is now head of the new School of Advanced Urban Studies at Bristol—related the Cublington decision to the whole corpus of planning policies in Britain since the 1930s. In particular, he saw it as a threat to the preservation of the green belts around London, to what he called London's 'open background', which performed the vital functions for the metropolis of providing scope for agriculture, forestry, recreation for city-dwellers, an alternative way of living for those who desired it, as well as accommodating a great variety of establishments, like hospitals and schools, which are better off in a rural rather than an urban setting.

'I believe . . . that the Vale of Aylesbury is a critically important part of this island. It is part of the fundamental hill and dale, forest and farmland break between London and Birmingham. It is of immense value to the nation . . . To locate the airport squarely athwart the break between the country's two largest conurbations, with the noise area extending from south-west to north-east for nearly forty miles, and with the consequent constraint on all the modest activities that the area can so conveniently accomodate in future, would seem to me to constitute nothing less than an environmental disaster.'[3]

Cublington's relevance to these broad issues of environmental concern in Britain is now apparent for all to see. It was also important in the context of a more specific matter of environmentalist activity—that is concern about the effects of modern airports. With their intense noise, their fumes, the way they dominate the landscape and throw their tentacles of link roads and airport services over a wide area, modern international airports represent a particularly serious blight on a large area of the environment. Professor Buchanan described them as 'hideously undesirable neighbours', an opinion which seems to be shared by people living near all of the world's busy airports. Airport resistance is certainly now a world-wide phenomenon. In the United States, there has been a series of amenity lawsuits against Los Angeles Airport, a national campaign to prevent a second Miami airport being built in the Florida Everglades and even Federal legislation which confines funds for new airport developments to those 'which shall provide for the protection and enhancement of the national resources and the quality of the environment of the nation'.[4] There are now signs of local unease with the development of the Third Paris Airport at Roissy-en-France, while in Japan the resistance to the Second Tokyo International Airport took the form of bloody, pitched battles between the police and a peasant–student coalition. The resistance at Cublington was never like that, nor was it likely to become so. It is worth noting that anti-airport groups from another Japanese city, Kobe, as well as from Norway and West Germany contacted the Cublington campaigners for advice. The Mayor of Kobe's delegation arrived in Buckinghamshire, expecting to hold conference in City Hall, Cublington: they eventually received the help they required at the home of the parish council chairman.

Cublington also occupies a special place in the development of anti-airport groups in Britain. It was the climax to the long saga of the third London airport, the culmination of resistance to the idea that it would ever be politically possible to build such a major airport on an inland site. The Cublington campaign thus has affinities—though there are also clear differences—with the earlier campaign at Stansted in Essex and with the campaigns at the other short-list sites considered by the Roskill Commission, Thurleigh in Bedfordshire, Nuthampstead in Hertfordshire and Foulness in Essex. Moreover, the arguments and methods used by the Cublington campaigners have not been without interest to amenity groups fighting to improve the quality of life in the shadow of Britain's existing airports. The support for the Buckinghamshire campaign from groups at Heathrow, Gatwick and Luton showed that.

Yet Cublington itself was not only an amenity issue and a campaign to conserve a pleasant stretch of natural environment. For 1,700 of its people at least it was a battle against the prospect of eviction from their homes and their land, eviction on a scale that is scarcely imaginable in Britain during peacetime. To these people, and particularly to the older and perhaps less adventurous ones, who had not known any other home than Cublington, the green fields and old churches meant less than the human environment of their families and neighbours. It was thus a community issue too, and the particular community threatened by the airport is an interesting one. Although Cublington is comparatively close to London and has in recent years become the home of a substantial number of professional people, it is in no sense an area of dormitory villages. It retains a strong rural identity and community consciousness. In some of the Cublington villages it is by no means uncommon to find households that are related to half a dozen other households in the same village and as many again in the village down the road. These networks of family relationships, taken together with the friendships and group activities of village life, give a rich texture to the communities of the area. This aspect of the Cublington story was noted by very few of the newspaper and television reporters who came from London to cover newsworthy events. Not surprisingly, they gave their attention to the articulate ex-urbanites who led the anti-airport resistance and ignored the ordinary country people who were equally opposed to the airport plan and equally active in resisting it, although both their motivation and their style of resistance may have been different from those of their middle-class neighbours.

There is, finally, another side to the Cublington story in the light that it sheds on the problems of government. The whole affair of the third London airport was, in fact, a tale of government by trial and error. At the time when Stansted was being considered as the airport site there was widespread resentment of the second-rate, bureaucratic planning methods which were revealed, and of the government's apparent readiness to ride rough-shod over the rights of local minorities. Stansted must be rated second only to the notorious Crichel Down affair as an example in modern times of almost every informed voice in the land being raised against the Government's handling of a planning decision. The Roskill Commission was, of course, set up to redress the grievances which had come to light in the Stansted affair. The 'Commission on the Third London Airport' was the most comprehensive and costly planning inquiry ever held in Britain: it cost the taxpayers £1,131,000 apart from its costs to other bodies. It called upon the services of a formidable army of talent, from its chairman, a High Court judge, and his six fellow-commissioners—each of national or international reputation in his field—to the many researchers, expert witnesses and lawyers ('the flower of the English planning bar', one observer called them), down to the humblest local authority clerk who spent two years amassing evidence for his council's submissions to the Commission. It also employed a tool of economic assessment, the so-called cost/benefit analysis model, which was unfamiliar to many people in Britain. The Commission's use of this tool led to a far-reaching debate about the implicit values of society as a whole, on matters like the monetary value to be put on old churches, the value of savings in travelling time, and the value to householders of the peace and quiet of their homes. This great debate, in which the people of Cublington and the other Roskill Commission sites played a full part, was by no means solved with the close of the Commission's hearings and the presentation of its final report. Yet, when the Commission did produce its final report, the Government at first failed to endorse it and then upset the site recommendation completely. One has to ask what sort of decision that was and why it was taken. Was it, as many people said, a political decision, taken in response to pressure inside and outside Parliament, including perhaps pressure from the people of Cublington? One Member of Parliament said later that 'these campaigners have been able to bring about the political overturn of all the principles of airline economics, regional planning and nature conservation through the political decision which they induced to locate a new intercontinental airport on a remote and unspoilt stretch of coastline, of rare and exceptionally wild beauty, in the south-eastern extremity of the country at Foulness'.[5] Was the Government's decision taken as a result of a radical change in its own thinking about airports and the environment? Or was the decision in any way an adverse judgment on the methods and results of the Roskill inquiry as such? These are questions that have to be pursued, for one of the stranger features of the Cublington story is that no Conservative minister since then has given a full policy explanation of why the popular, but still controversial, decision to build the third London airport at Foulness was taken.

These weighty questions of government will inevitably take this book some distance from the popular demonstrations and the newspaper reports of the Cublington villagers. But the deliberations of the Roskill Commission often seemed equally remote to people whose homes and lives were being weighed in the balance. The people of Cublington make their first real appearance in the book at Chapter V and the narrative of their anti-airport campaign is related in the subsequent chapters. Chapter IV describes the setting up of the Roskill Commission and its early work before it produced a short list. Chapter III is about the Stansted Affair which led to the creation of the Roskill Commission. Chapter II tells how the present controversy and conflict of interests over airports in Britain have arisen.

II
THE AIRPORTS PROBLEM

The feature of modern airports that makes them so undesirable as neighbours is the noise they generate. In south-east England, which the Roskill Commission described as 'the Clapham Junction of the air', there are between two and three million people who to some degree suffer from the effects of aircraft noise. For them the distinctive roar and high-pitched whine of jet airliners landing and taking off from Heathrow, Gatwick, Luton, Stansted or Southend forms the daily background to their lives. It is a peculiarly oppressive sound which has to be lived with to be truly resented. People unfamiliar with it may be surprised to read that prolonged exposure to this noise can retard the mental development of children and, in adults, can cause mental illness and sexual impotence and even shorten the expectation of life. There is a reported case of a girl student who became so disturbed by jet noise that she suffered spells of hysterical blindness. Yet these stories serve only to illustrate what two million people in southern England and many more elsewhere know for a fact, that living near a major airport is no longer a tolerable existence. 'Noise slums' is the phrase that Mr Michael Noble, then President of the Board of Trade, used to describe conditions near Heathrow Airport in 1970.

When the people of North Buckinghamshire first learned that the Roskill Commission was considering Cublington as a site for the third London airport, very few of them grasped the fact that their villages and towns could become 'noise slums'. Not surprisingly the inhabitants of the on-site villages were quicker to appreciate that they were in danger of eviction, although there was initially a serious dearth of official information about the exact area of the site. But for the rest of the people of Cublington the prospect of living in close proximity to an airport larger and even busier than Heathrow was something beyond the range of their experience. Would aircraft noise near the airport be a merely distant drone that was irritating yet tolerable? Would it interfere with normal activities like conversation or school work? Could the nuisance be alleviated by sound-proofing? Would compensation be available for noise nuisance? How did one measure noise nuisance and what were 'noise contours'? One of the first tasks of the Wing Airport Resistance Association was to provide the answers to these questions and, by a series of evening meetings in village after village, to create an awareness of this unfamiliar world that Cublington was going into.

Measuring aircraft noise is a highly technical affair,[1] but it is a discipline that has had to be learned by the neighbours of existing airports if they are to make their complaints to the authorities. Noise in general is measured in either decibels (dB) or 'perceived noise decibels' (PNdB). The nuisance of aircraft noise, however, is measured in Britain on a different scale called the 'Noise and Number Index' (NNI). This takes account of both the level of noise and the number of times that it is heard. Thus, 40 NNI can represent either thirty flights a day each generating 97 PNdB or sixty flights a day at 92 PNdB, which is half as loud (the decibel scale is logarithmic). The Noise and Number Index was devised by Sir Alan Wilson's Committee on the Problem of Noise, which reported to the Government in 1963, and was based upon an opinion survey of householders' reactions to noise near Heathrow.[2] The Wilson Committee plotted on a map the points at which various intensities of noise were experienced and then joined the points of equal value together to form 'noise contours'. The noise map of an average major airport looks somewhat like the outline of a dolphin, with its nose pointing along the take-off paths and the lateral fins and tail representing the spread of noise from the take-off and landing points. The noise contours devised by the Wilson Committee corresponded roughly with the degree of local reaction to the noise. People living within the 60 NNI contour, that is very close to the airport perimeter or right under the take-off path, found aircraft noise 'intolerable', those living between the 50 and 55 NNI contours were beginning to find it 'unacceptable', those living between the 40 and 45 NNI contours found it 'moderate' to 'difficult' and those living on the 35 NNI contour found the noise on the whole 'tolerable'. These were average reactions, of course, based upon the Heathrow survey. The Noise and Number Index has obvious deficiencies as an indicator of the aircraft noise tolerance of the population at large, but no one has yet found a better indicator. However, authorities in other countries tend to put the 'unacceptable' level of aircraft noise much farther away from the airport perimeter than did the Wilson committee, in fact closer to 40 NNI than 55 NNI. Almost 17,000 households at Cublington would have come within the 40 NNI contour, excluding the thousand displaced households on the airport site itself.

Once they had begun conjuring with these technicalities, the people of Cublington found themselves in the company of a nationwide anti-airport movement. To a large extent they were able to make common cause with each other and they received and gave much help and advice, but there was at times a shadow of potential rivalry hanging over them. This element of divisiveness is always present in environmental campaigning, especially where one group's airport victory can mean an increased burden of noise for another group.

Cublington had joined the new world of airport protest. Chris Brasher, writing in the *Observer*, addressed this cautionary tale to the people of Cublington, Thurleigh and Nuthampstead. He called it 'Wait Till the Jumbos Get You':

'We first had to learn how to complain. It is no good just ringing up London Airport (No 1 in our case) and saying: "You're making a devil of a noise." You must be specific. It was only the other day that we discovered that we had all developed the same technique. As soon as you've recovered from the blast, you rush to the phone and dial TIM. That gives you the accurate time of the nuisance. Then you dial Skyport 4321 and say you want to make a noise complaint. They really are charming but don't be diverted. Give them the time of the plane and ask for its identity, its noise level and its height. You won't get all that information. Indeed, we've never yet got the height of a plane from the authorities. In all their literature they say it's a vital factor but they don't bother to check on it.

'What you will get is a charming letter back from the British Airports Authority identifying the airline and saying that this plane did not infringe the noise regulations. And they add their profuse apologies. Seeing that you have just been blasted out of bed with all the windows rattling and the children crying, you deduce that there must be something wrong with the noise regulations. There is. They are a farce.

'When you press your complaint you are referred to the Board of Trade Civil Aviation Department, telephone number HAYes 6171—you are going to need that number, it will save you a lot of letter writing. We won't bore you by taking you through the voluminous file that we have amassed because it is full of soft soap and it takes many letters and complaints to get at the facts.'[3]

Many a peaceable suburban man has been changed into a vengeful goad of Authority by the sort of bureaucratic frustrations described by Chris Brasher. In the last three years, moreover, the position has worsened. Responsibility for noise from civil aircraft is now divided between the Department of Trade and Industry, the new Civil Aviation Authority, the British Airports Authority and, to some extent, the Department of the Environment. Even the officials themselves find it difficult at times to pin down just who deals with a particular noise complaint.

If there are such obstacles to the ordinary citizen's obtaining even the facts, then his chances of alleviating the problem by his own efforts or those of his local amenity organization are small indeed. Moreover, the fact that jet noise did not assault south-eastern England in one rush—it was first felt at Heathrow in 1958, at Gatwick some seven years later and at other airports later still when the boom in charter traffic began—has meant that airline and airport operators have been able to play off one amenity group against another, by switching airlanes and stacking areas. In these circumstances the creation of a national lobby against aircraft noise became inevitable. The first organisations in the field were the old-established Noise Abatement Society and the British Association for the Control of Aircraft Noise (BACAN). Later, two more militant bodies were formed: the United Kingdom Federation against Aircraft Nuisance (UKFAAN) and the Local Authorities Aircraft Noise Council, which co-ordinates the anti-noise campaigns of councils mainly in West London and the Heathrow area. Together these latter bodies have mobilised an effective parliamentary lobby and formed links with anti-noise groups in Western Europe. There is also a Four Airports Group, which embraces local amenity groups at the four state-owned airports or airport sites in the London region—Heathrow, Gatwick, Stansted and Foulness.

But the difficulties faced by this lobby are not just technical and bureaucratic. As Mr Justice Roskill explained in the Commission's final Report:

'Those who suffer from aircraft noise, though able to exert powerful pressure through many bodies, are effectively denied legal redress. Section 9 (1) of the Air Navigation Act 1920 first denied recourse to the courts for trespass by a nuisance caused by aircraft flying over property and the Civil Aviation Act 1949 by Section 40 (1) substantially re-enacted the provisions of the earlier statute. The present compensation laws help only those whose homes are expropriated. Those whose homes are rendered almost uninhabitable by noise receive no help save in some cases a contribution towards the cost of partial sound-proofing; they have either to stay and endure the noise or to incur the costs and disruptions of moving . . . The frustration and sense of injustice felt by those on whom the noise burden is inflicted add a powerful emotional element to discussion of the problem.'[4]

The crux of the problem, as many people see it, is how to make modern jet aircraft learn to live with people. The Roskill Commission, although for national economic reasons it recommended that the airport should be built at Cublington, also urged that the compensation laws of Britain should be extended to those 'whose homes are rendered almost uninhabitable by noise'. This would have given some small financial help at Cublington to people living in villages like Soulbury, Whitchurch and Wing. Not surprisingly they found little comfort in this small concession and fought the more strongly for the airport to be built away from people altogether at Foulness. The Government agreed with them. It is also worth noting that in 1972 UKFAAN was seeking a ruling from the European Commission on Human Rights about the denial of legal redress to aircraft noise sufferers in Britain. But these were only moves on the very edge of the problem of how to make modern air travel compatible with the lives of the people on the ground. Since publication of the Roskill Report, the Conservative Government has taken Britain's first major step in this direction (partly in response to the pressure from Cublington) by deciding to build the third London airport on the Maplin Sands, rather than at an inland site. 'This is the world's first environmental airport, of which we shall be proud,' Mr Eldon Griffiths, Under-Secretary for the Environment, told the Commons a year after the Foulness decision.[5] The Conservatives had already taken this concern for the environmental problems of airports a stage further by decreeeing in the summer of 1971 that, when Foulness (or Maplin) comes into operation in the 1980s, both Stansted and Luton will be run down and noisier jets will be forced away from Heathrow and Gatwick to the new airport. This however is a long-term strategy. The same can be said of the often-expressed hope that the noise problem will disappear when 'quieter' jets come into service, either through the introduction of new types of aircraft, like the Lockheed Tristar with its Rolls-Royce RB 211 engines, or by the fitting of 'hush' kits to the engines of present-day aircraft. It was the considered opinion of the Roskill Commission, however, that the continued use of present-day aircraft and the increased weight of new aircraft like the Jumbo jets would produce a net *increase* in noise levels up to the 1980s. Not until 1985, said the Commission, could one expect the noise levels of aircraft to drop to the level they were at in 1967 and one could not reasonably hope for any real alleviation of the problem before the end of the century.[6]

In the short term, the noise problem will clearly get worse. That is why the anti-noise lobby has not in any way relaxed its campaign since the Conservative Government revealed its environmental strategy for airports. Some of the measures that are being urged on the Government are: a ban on night flights from all urban airports; zoning of land around new airports to prevent the creation of further 'noise slums'; preferential landing fees for 'quiet' aircraft; expansion of the very limited grants for sound-proofing homes afflicted by noise (possibly to be paid for by the preferential landing fees); and research into different procedures for landing and take-off. One of the few noise-alleviation measures of which pilots have to take note is the requirement that planes taking off from Heathrow must cut back power at about 1,000 feet and climb more gradually until they have cleared the built-up area. It is an unpopular procedure both among environmentalists, who argue that it spreads the noise over a wider area, and among pilots who are worried about the safety factor. This worry became all too apparent after the BEA Trident crash at Staines in June 1972. Consequently, UKFAAN has submitted to the Department of Trade and Industry a plan for 'steep climb' and for 'two-angle descent' at Heathrow. The plan was drawn up as a result of tests carried out on behalf of UKFAAN by the pilots of several major airlines. The pilots carried out these tests without the knowledge of their employers; they are, indeed, part of the anti-noise 'fifth column' that is at work not only in the aviation industry, but also in the Civil Service and government research establishments.

There is one major obstacle to the implementation of an environmental strategy for airports in Britain. It is the lack of any national plan for airports and any central ownership or control. The position at the moment is chaotic. The state corporation known as the British Airports Authority controls only five airports—Heathrow, Gatwick and Stansted in the London region, and Prestwick and Edinburgh (Turnhouse) in Scotland. In time it will control Maplin. All other airports in the country (and there are a considerable number of them) are owned by other bodies. It is a fact, of course, that the B A A does control Britain's busiest airports, but there is no guarantee that this will always be the case, especially since airports in the industrial Midlands and North-West are likely to gain traffic when the remote Maplin airport on the east coast comes into operation.

The process by which Britain arrived at this chaotic system of airport ownership is an interesting one. At the end of the Second World War, there were many chances for making a new start—new types of aircraft, new routes, new techniques like radar, new airports and new airport installations like concrete runways and permanent hangars. It was the ideal time to draw up a national airports policy and to reserve the land for the airports of the future. During the war some 625 airfields had been put into service for the R A F and the U S A A F, 444 of them with concrete runways, a concreted area altogether of 36,000 acres—the equivalent of a city between the size of Edinburgh and Birmingham.[7] After the war much of the concrete was dug up, but many airfields were not fully returned to their pre-war state. Even airfields which had again come under the plough often remained on some file in Whitehall marked as former airfields which could again be used for aviation. They became environmental time-bombs ticking away into the future. One has only to look at an Ordnance Survey map of Southern England, the Midlands or East Anglia to see how many 'former airfields' are still marked as such. Many of these sites were scanned half a dozen times in the search for the third London airport. There are 'former airfields' at both Cublington and Nuthampstead and an existing private airfield at Thurleigh. Britain in the Second World War was really an aircraft carrier, said our enemies. The Roskill Commission repeated this 'hostile jibe' with satisfaction and said that it should 'in the last thirty years of the present century be a source not only of pride but of economic and political sense'.

In 1945, the Labour government published a White Paper, which promised that all the airports needed for scheduled services would be brought under the control of the new Ministry of Civil Aviation. In 1947, the Government got round to publishing a list of 44 airports that it planned to acquire. But after 1947, as more pressing problems of the Age of Austerity beset that administration, there was no longer serious talk about a national airports plan. Instead, the Government concentrated on the airports already under its control, and especially on Heathrow which had been marked out as a major airport but which had very primitive terminal facilities before 1948. This was not an age in which airports were money-spinners. As late as 1955, two years after Heathrow had taken over from Northolt as Britain's busiest civil airport, the Ministry was still showing a loss of £6 million on its airport account.

The 'Have-a-Go' policy of airport ownership (as Professor Buchanan described it) really got going after 1951, when the Tories returned to power. Private airlines began to operate in growing competition with the state corporations and wherever possible they flew out of different airports from their nationalised competitors. The Government began the policy of selling off, rather than buying, airports. Virtually any organisation, council or individual who thought himself capable of operating an airport could now do so—provided of course that he had the capital and could satisfy the Ministry's not too onerous requirements about lack of physical obstacles to flying and the provision of fire-fighting facilities. Questions of planning and amenity seldom arose. As Rigas Doganis wrote in his pamphlet *A National Airports Plan*,[8] 'as all pretences of planning were progressively abandoned, the system of ownership and management of airports became increasingly confused and complex. In 1945 the Labour government promised order. Today there is chaos.'

Mr Doganis was writing in 1967. In the 1970s the situation is if anything worse. Of the forty-five or so all-weather airports used for civil aviation in Britain, over half are owned by local authorities. Some are wholly owned by one council, like Luton, Southend, Glasgow and Manchester airports; some by two councils like Leeds/Bradford and Gloucester/Cheltenham airports; and some by consortia, like the East Midlands Airport at Castle Donington which is owned by Derbyshire, Nottinghamshire and Leicestershire County Councils and the town councils of Derby and Nottingham. Some municipal airports are run by private companies acting as the councils' agents, like Exeter, Plymouth and Southampton. Also in the local authority ownership class are the Channel Islands and Isle of Man airports, but those which serve the Isles of Scotland are run by the new Civil Aviation Authority set up in 1972, as are the larger airports at Aberdeen, Inverness and Belfast. The airport on the Scilly Isles is run by a commercial company, BEA Helicopters. Another group of airports is privately owned, either by airlines or aircraft manufacturers. And of course the British Airports Authority controls its five airports.

There are other anomalies. In central Scotland, for example, Prestwick and Edinburgh are run by the BAA but the third airport in the

Edinburgh–Glasgow region, at Abbotsinch, is controlled by the city fathers of Glasgow. There is no all-weather airport serving East Anglia or North Wales, yet Hampshire has three airports in close proximity—Bournemouth (Hurn), Southampton (Eastleigh) and Portsmouth. In the south-east, Luton and Southend, which as has been noted are municipally run, compete for the seasonal holiday traffic of the London region with Gatwick and Stansted, which are state airports.

'Such a complex and diverse system of ownership,' wrote Mr Doganis, 'cannot but create confusion and local rivalries. There is no machinery for co-ordination, no plan or framework within which separate airport authorities can plan their own development in the knowledge that their forecasts and expectations will not be upset by unexpected developments at neighbouring airports. On the contrary, airport owners enjoy almost complete freedom of action, especially if they do not need government aid. To attract traffic upon which their revenue and prestige depend, they try to outdo each other in the facilities they provide. This is both costly and wasteful.'

One of the questions that one comes back to again and again is who carries the ultimate responsibility for airport planning in Britain. Some interesting light was thrown on this in the evidence taken by the House of Commons Select Committee on the Nationalised Industries when it studied the British Airports Authority in 1970–71. The British Airports Authority itself told the MPs on the Committee that it would have liked to see a properly thought-out national plan long ago, but it was not possible for the Authority itself to draw one up. Would the Department of Trade and Industry be in a position to formulate a national plan? 'No,' said the Department's senior civil servant in charge of aviation, 'we have not got the numbers, we have not got the money, we have not got the experts available to put on this task in a high, wide and handsome way.' The Roskill Commission had these resources, the Under-Secretary added. 'I think everyone hoped two and a half years ago that the Roskill Commission would in effect plan a very great part of our aviation infrastructure, that is the south-east and the area adjoining the south-east.' But surely, said an MP on the Select Committee, the Roskill Commission was not asked to draw up a national plan. 'Absolutely,' agreed the Under-Secretary. Who then drew up the terms of reference for the Roskill Commission? The Department of Trade and Industry itself. The MPs did not leave it there. Would anybody be attempting to formulate this long-needed national plan for the airports of the future? Yes, said the Under-Secretary, the new Civil Aviation Authority which was to take over the Department of Trade and Industry's aviation duties—but too much should not be expected from the new Authority. 'I hope that in emphasising the point that a plan is not a panacea for all the ills of airport development, I did not move to the other extreme and suggest that there was no point at all in preparing a plan. This is a job we shall ask the Civil Aviation Authority to embark upon, but it is a very slow, long and costly process especially as there must be, I think everyone would agree that this is necessary, public participation both at local and at national levels . . . one does run into, and rightly, public opposition, or indeed disappointment that they have not got a bigger circle on the plan. But certainly I think the Civil Aviation Authority will be charged with the job of preparing what I call an indicative plan.'[9]

Since 1971, there has been little progress towards a national airports plan. In July 1972, a Bradford MP, Mr John Wilkinson, initiated a Commons adjournment debate on the demand for such a plan (his interest was not so much in environmental considerations as in bringing more air traffic to Yorkshire). The answer that Mr Wilkinson received from a junior Minister of the Department of Trade and Industry was: 'We share the desire to see an effective national airports strategy . . . although it is not necessarily my view that all will be solved at one stroke by the provision of a national plan.'[10] The Government's 'strategy' is that it is prepared to use its powers of veto on planning applications to restrain development of environmentally bad airports, like Luton or indeed Bradford. The difference between this strategy and a national plan is that the former is unrevealed until tested by individual planning applications. The lack of a plan still leaves everyone concerned—airlines, local authorities, airport users, business, travel agents, workers and environmentalists—in the position of having to fight each battle as it arises without relevance to the battle being fought in the neighbouring airport area.

One last point is worth drawing out from this rapid resumé of the ills of Britain's airports system. It concerns the public servants who operate the system, both in the civil service and at the airports themselves. Almost without exception, the one thing that strikes every member of the public who had dealings with these public servants is their almost Byzantine lack of frankness. It comes to light in efforts to register complaints about aircraft noise, as Chris Brasher found. It comes to light in decisions made by the aviation establishment, as when the people of the Gatwick area were told that their long-demanded limit on night flying would be implemented—only to find later that whereas the limit already imposed at Heathrow ran from 11.30 p.m. to 6 a.m., the Gatwick limit would operate from midnight to 6 a.m. and in the extra half hour Gatwick might be handling as much as 10 per cent of its whole night traffic. The Select Committee, too, commented on this lack of frankness and made it the subject of a formal recommendation: 'accepting the commercial and other risks, a fuller, franker and earlier setting forth of the Authority's plans, aspirations and problems would evoke correspondingly greater co-operation and help from employees, airline operators and other users, and the surrounding communities'.

There may be many good reasons for this secretive attitude among the aviation establishment. Airport services have suffered from a chronic lack of public funds. The Civil Servants in charge of them have held the conflicting responsibilities of making a profit, ensuring air safety and serving the interests of the local communities. They have often had to go cap in hand to foreign airlines in a business which is becoming increasingly competitive between different countries. And, of course, they have faced the organised wrath of the environmentalists. The public servants have felt harassed and they have said so. But their lack of frankness has become almost an occupational characteristic of their breed.

The people who felt this most strongly were the residents of the Stansted area. The story of Stansted is indeed the story of one secretive move after another, until the whole thing became so untenable that the Roskill Commission was appointed to decide the issue in the full light of public proceedings.

III
THE STANSTED AFFAIR

The people of Stansted and the people of Cublington fought the same enemy. It was Authority and its attempt to deprive them of their homes, their farms, their rural tranquillity so that it could build a monstrous airport in their place. In retrospect, one can see that one battle followed naturally on the other and that both Stansted and Cublington fitted into a broader campaign to defend the green environment on London's doorstep. The differences between the two battles, though, were also striking. Cublington's tactics were more aggressive, more resourceful, more consciously aimed at public opinion. There were also, it seems, underlying differences in their objectives. Stansted was saying: 'All we want is a fair hearing and proper regard for the basic rules of justice—if we get that and the result still goes against us then, of course, we shall accept our defeat.' Whether they would have done so is perhaps another matter, but at their stage of the game the Stansted people were seeking only justice. At Cublington, on the other hand, the voice of the majority said: 'Nobody in his right mind would want to put an airport here—in fact, nobody will put it here because we shall not let them, and we mean it.' That last bit admittedly worried some of the Cublington leaders who were conducting the campaign on a legal and political level.

Cublington's stand came of course four years after the Stansted affair, when the environmental issue seemed to many people in Britain and elsewhere to have reached a crucial, even desperate stage. There had also been a fundamental change in the tactics of the enemy they were both facing. When Stansted went out to do battle, it faced a government and civil aviation establishment that was obstinately defending a plan that most people could see was indefensible. It had been demolished at a public inquiry and vindicated only at a subsequent inquiry held in secret. Even Mr Justice Roskill could not refrain from commenting on the parlous conduct of public affairs that led to the establishment of his own Commission.

'No member of the Commission,' he wrote, 'embarking upon the formidable task entrusted to him and his colleagues could have failed, as an ordinary citizen and before his appointment to the Commission, to be aware of the deep feeling which the decision in the White Paper [of May 1967] had aroused. During the years preceding the Commission's appointment the "Stansted issue", as it came to be known, was widely canvassed both within and without Parliament. Intense local opposition was organised, charges of bad faith, of central government riding rough-shod over local government and the rights of minorities were freely made . . . It was argued that full weight had even then not been given to factors such as the social costs and the planning and transport problems ancillary to the development of a new major airport.'[1]

Mr Justice Roskill's purpose here is plain. He is not only relating the political background to the setting up of his own Commission on the Third London Airport; he is also marking out his methods of inquiry—his weighing up of the social costs and planning and transport problems—as being very different from the methods used in the past. It is a legitimate claim. But from the environmentalist point of view, the Roskill Commission looked suspiciously like the old enemy in a new guise. It was as if the Philistines (the environmental Philistines) had learned their lesson from Goliath's defeat by a lad with a sling and found a new champion whose armour they had redesigned to give him more mobility and make him less vulnerable around the brain. But when they sent him out to battle, they found a new and fiercer David waiting for him.

The background to the Stansted affair can be simply laid out. The airport lies to the east of the town of Bishop's Stortford, about 35 miles north-east of London. It was built in 1942 for the United States Army Air Force and reoccupied by them after the war when they extended the runway to 10,000 feet, making it one of the longest in Europe at the time. When the Americans left for the second time, it became the property of the Ministry of Civil Aviation who used it mainly for training air crews. In 1953, it was shortlisted—along with Gatwick, Blackbushe in Hampshire and Dunsfold in Surrey—as a possible alternative airport for Heathrow but rejected because it was thought to be on the 'wrong side' of London. In 1960, the Ministry told the House of Commons Estimates Committee that Stansted was making a loss, whereupon the Committee urged 'an immediate, detailed study of the prospects of Stansted as a future third airport for London'.[2] The Committee added its own observation that, as a potential London airport, Stansted had poor rail and road links with the capital and raised difficulties for neighbouring civil military airfields.

So in 1961, the Minister of Aviation (by this time he was no longer 'Civil') set up the famous Interdepartmental Committee on the Third London Airport. Its chairman was Mr George Vinger Hole, Under-Secretary of the Ministry's Aerodromes Division, and twelve of its other fourteen members were also drawn from the civil aviation establishment, either as civil servants or officers of the state airlines. The two remaining members were from the Ministry of Transport and the Ministry of Housing and Local Government. This committee began its work with a sweep through a hundred former airfields or possible ones 'within an hour of central London'. It then narrowed these down to a short list of eighteen. Of these, three were active military airports and three were ruled out for other reasons. That left twelve sites—two to the west of London, one to the north and nine to the east. This is an interesting geographical balance when one recalls the categorical statements of the Roskill Commission and the British Airports Authority that people to the east of London should not have a major airport because they do not show a 'propensity to fly'. The Interdepartmental Committee's remaining dozen included three municipal airports—at Luton, Southend and Hurn—and three new coastal sites at Sheppey, Cliffe and Foulness, which were presumed to be ruled out by their proximity to the Shoeburyness Firing Range. So in June 1963, the Committee reported back and recommended Stansted. The Interdepartmental Committee's report[3] was published in March 1964 with a note from the Conservative Minister of Aviation, Mr Julian Amery, endorsing the choice.

'Though not perfect it seems to be the only suitable site,' he wrote, adding the prophetic note that 'the choice of a site for a new airport will not please everyone. Wherever it is put, it will take away land that could be used for other important purposes. There are those who viewing the choice from the aspect of amenity alone would like to see it anywhere but in their own neighbourhood. Others will welcome it as near as possible to their factories or offices. The choice is limited by technical considerations. The report discusses these in detail. It concludes that Stansted airport should be selected and designated as London's third Airport. The Government believes that this is the right choice.'

At the end of his foreword, Mr Amery added the pious thought: 'I hope that there will be full public discussion of this report and I shall welcome constructive suggestions for making Stansted an efficient and attractive airport.' Whatever constructive suggestions the Minister received, they were not made public.

The report itself was made public in March 1964. Its full implications were not known until July when the Ministry convened a meeting of local authorities at Harlow where a map was produced. It showed a development covering 19 square miles, with two pairs of parallel runways aligned so that aircraft would take off and land either over Bishop's Stortford or over the town of Great Dunmow. The map was promptly published in the local weekly paper, *The Herts and Essex Observer*, where its effect was electric. As Olive Cook put it in her paperback, *The Stansted Affair*: 'At this stage the details of the plan were no more than proposals but the news shocked the people of Essex and Hertfordshire and they registered their alarm at hastily-called public meetings and in showers of letters to the Local Press.'[4]

The most significant of these meetings was held in the village hall of Takeley, one of the two communities on the edge of the planned Stansted expansion. It was scheduled as a meeting of the parish council but the hall was packed to overflowing (a scene that was to be repeated five years later at Stewkley and Wing). One reason for the large Takeley gathering was the presence of the local Member of Parliament, Mr R. A. Butler (now Lord Butler of Saffron Walden) who was also Foreign Secretary. Mr Butler read to the meeting a letter from the Prime Minister, Sir Alec Douglas-Home, who promised that a public inquiry would be held in the following year. 'It is important,' said the Prime Minister, 'that this inquiry should be thorough and genuine, and I want to make it clear that the Government has in no way made its final decision, and cannot do so until it receives the report of the inquiry.' Sir Alec's words were to echo with a hollow ring throughout the Stansted affair of the next three years.

The main result of the Takeley meeting was the setting up of a local resistance association, called at first the North-West Essex Preservation Association (East Herts was added later to this already cumbersome title). It was to be an 'umbrella' organisation, with local committees and other groups affiliated to it. By the time that the public inquiry opened in December 1965 at Chelmsford, forty village and town groups were associated with it (going under titles like 'Preservation Association', 'Anti-Airport Committee' and even 'Fighting Committee') and thirty other groups affiliated (ranging from the local hunts and beagles to a cluster of East London cycling clubs). The impetus for the formation of the NWE&EHPA had come from a small band of farmers and residents. The joint-chairmen were Sir Roger Hawkey, Bart, a businessman living at Takeley, and Mr John Lukies, a farmer and magistrate of Great Dunmow. A seventeen member executive committee was formed and met once a fortnight; a council of representatives from associated village committees met about once a month. A retired merchant banker become the Association's full-time general organiser and there were three full-time secretarial workers. The Association distributed a monthly newsletter, but the first objective was not publicity but the raising of funds for the local inquiry to be held at Chelmsford. Altogether £23,500 was raised for this purpose, in what Sir Roger Hawkey describes as 'small donations from ordinary people'. There were 13,000 subscribers and only three were said to have given more than £500 each. Later in the campaign, a further £15,000 was raised to continue the fight.

The next objective was the mobilisation of public opinion in the area. Supporters were urged through the newsletter to 'write to the Prime Minister, Cabinet Ministers, including the President of the Board of Trade, Members of Parliament, Lords—write to the newspapers declaring your opposition to the "crime of Stansted" '. There were also car-stickers—'NO to the Third London Airport at Stansted'—and slogans—'People Before Plans' and 'Stop Jay Walking over Stansted' (Douglas Jay, the Labour President of the Board of Trade from 1964 to 1967 was to become one of the prime villains of the piece).

During the early stages, at least, the Association made little attempt to influence or attract the communications media except through individual letters to editors. At one time there was discussion about hiring a PR firm, but there were found to be insufficient funds. At a much later stage, however, the Association found its own PRO in Mrs Susan Forsyth, the daughter of one of the co-chairmen, Mr Lukies; she commuted daily from Stansted to Fleet Street where single-handedly she wooed the Press. But in the main the North-West Essex and East Herts Preservation Association eschewed any obvious appeal to the news appetites of Fleet Street or television. At no time did they appear ready to back their protests with sanctions of the sort that would have brought the press running. As Olive Cook observes:

'The members of this organisation were activated in the first place by their keen awareness of what was at stake in the locality, by their desire to help all those whose lives would be disrupted by the project and by their conviction that it would not be in the national interest to place a major international airport in a region where access to London was already congested and where the large-scale development necessitated by an airport could not be reconciled with the demands of regional planning. But during the past three years, and especially during the past six months of the present year (1967), these men have become far more than the leaders of local opposition: they have emerged as the champions of all fair-minded people against perverse and obdurate bureaucracy, of the rights of the individual against the machine. Although, belonging to the threatened area as they do, they had good personal reasons for opposing the Government's proposals, they have throughout maintained their willingness to accept the choice of Stansted as the site for London's Third Airport if that choice could be shown to be the best possible.'[5]

Olive Cook's paperback is the best account there is of the Stansted resistance, and a true reflection of what many local people felt. If it errs, it does so by giving too much credit to the NWE&EHPA and not enough to other anti-airport forces. The Association certainly had powerful friends. During the first year, Essex County Council threw its weight behind the anti-airport campaign and was later to spend tens of thousands of pounds on resisting the Government's plans, even as far as a High Court action against the Government. Hertfordshire County Council also came into the campaign along with the Dioceses of Chelmsford and St Albans: the Chelmsford Diocese contributed to the Association's funds and asked to be represented by the Association's counsel at the public inquiry. Among individual friends, the Association could number Sir John Elliot, former chairman of London Transport, a director of BEA, and a member of the British Airports Authority when it was established in 1966; Sir John was on the Association's executive committee and also chaired the Essex Residents' Association, a smaller environmentalist pressure goup. Another local resident was the industrialist and government adviser, Lord Plowden, who was later instrumental in bringing together the Stansted objectors and the Conservative peers in the House of Lords in what was later to be a crucial alliance.

After the organisation of local resistance came the long-awaited public inquiry. It was held in County Hall, Chelmsford, from December 6 1965 to February 11 1966—a total of thirty-one working days—before Mr G. D. Blake, an Inspector of the Ministry of Housing and Local Government, and his Technical Assessor, Mr J. W. S. Brancker. Mr Brancker, who later became a prominent ally of the Stansted objectors, was an aviation authority in his own right as a former Deputy Assistant Director of the International Air Transport Association (IATA) and the son of Air Vice-Marshal Sir Sefton Brancker, the first Director of Civil Aviation in Britain. There was something extraordinary about the presence of such an eminent assessor at what seemed to many people to be an ordinary planning inquiry. Its terms of reference after all were: 'to hear and report on local objections relating to the suitability of the choice of Stansted for an airport and the effect of the proposed developments on local interests'. In fact, it was a most extraordinary inquiry. It was non-statutory, since the Minister of Housing and Local Government was under no statutory obligation to hold an inquiry into the development of Crown Lands.[6] The Minister would not have been obliged to accept the findings of his Inspector even if it had been a statutory inquiry. On the other hand, there were a number of reasons why the inquiry should be not only 'thorough and genuine', as Sir Alec Douglas-Home had promised, but also involve a 'full public discussion' of the many issues, as Mr Amery had promised. It was a planning question of much more than local importance, and it had already led to vigorous political activity in the Stansted area and the country at large. The Preservation Association was thus able to gain from the new Labour Government's Minister of Aviation, Roy Jenkins, an assurance that 'it will be open to objectors to suggest modifications to the outline scheme of development, or to propose alternatives, but not to question the need to provide a third major airport to serve London'.

The objectors used to the full their freedom to suggest alternatives, but first they set about demolishing the case for Stansted. The Government's case was badly prepared. The aviation establishment had not worked out in advance whether the expanded operation of Stansted was compatible with the continued use of the Wethersfield USAF base. It was not sure how many homes would be affected by aircraft noise from Stansted nor the value of the agricultural land that would be seized. It had not prepared a defence to the argument that Stansted's expansion conflicted with all official views and priorities about regional planning. Finally, under cross-examination a government witness agreed that Stansted was between eighty-five and ninety-five minutes' journey from London, and a move was made to wind up the inquiry on the grounds that the basic criterion of the airport being within an hour of central London had been abandoned. The Inspector, Mr Blake, would not allow this, and the objectors then concentrated on suggestions of alternative sites. There was considerable support for Padworth on the borders of Hampshire and Berkshire and for Sheppey and Cliffe on the Kent side of the Thames Estuary. The Government raised air traffic control objections to all of these, though Mr Brancker, the assessor at the inquiry, was later to state that the existing air traffic control patterns were not immutable. Foulness was first suggested by the chairman of the Noise Abatement Society, Mr John Connell, who produced a plan for a twenty-minute mono-rail service between Foulness and London. Foulness and Sheppey, however, fell foul of the Shoeburyness Firing Range and the Government produced at the inquiry the Director-General of Artillery, Major-General Egerton, to say there were no alternative sites for it. Yet over and above all these counter-arguments and alternatives, the Chelmsford inquiry showed two things: that a local inquiry was clearly not the right forum in which to decide the location of a major international airport, and that a local amenity group, well-funded, well-organised and with expert counsel (Mr Peter Boydell, QC) could seriously challenge a Whitehall plan in open debate. As R. E. Wraith and G. B. Lamb wrote in their book, *Public Inquiries as an Instrument of Government*:

'The inquiry provided a classic example of the difficulty of distinguishing between "an objection and the thing objected to", for the "local objections", as developed by Counsel, proved in effect to be a comprehensive attack on certain fundamentals of the Government's proposal.'[7]

The inquiry closed in February and Mr Blake presented his report to the Government in the summer. It was not to be published for another year, but in it the Inspector said that the Stansted plan succeeded only on the grounds of air traffic viability. There were 'formidable and justified' objections against it, he said, on the grounds of planning, road and rail access to London, noise, local amenity and the value of agricultural land to be taken. 'It would be a calamity for the neighbourhood,' he concluded, 'if a major airport were placed at Stansted. Such a decision would only be justified by national necessity. Necessity was not proved by evidence at this inquiry.' When it received the report, the Government immediately carried out what it said was 'a comprehensive and searching re-examination of the many complex issues'. The methods and the names of officials carrying out this re-examination were never made public, although it was strongly suspected at Stansted that some members of the discredited Interdepartmental Committee had taken part in it. In the Stansted White Paper, eventually published in May 1967, three points about this re-examination emerged. The first was that nine alternative sites had been examined and found unsuitable—Castle Donington (Derby), Ferrybridge (York), Gunfleet Sands, Dengie Flats and Foulness (all off the Essex coast), Sheppey, Cliffe and Plumstead marches (all on the south bank of the Thames) and Padworth. The second was that the area north-west of London had been found to have a 'particular attractiveness from a regional planning aspect'. Consequently two representative sites—at Silverstone and Thurleigh—had been examined and rejected on the grounds of interference with military aviation and poor access to London. Finally, it was revealed, Stansted had been chosen yet again. The third London airport would be built there—but not with three or four runways but six. There would be three pairs of parallel runways and perhaps more if they were not parallel.

The White Paper was published on Friday, May 17, the day before Parliament recessed for Whitsun. It was received with outrage, both at Stansted and more widely. The press was overwhelmingly against it, describing the White Paper and Douglas Jay's defence of it as 'wrong' and 'depressing' (*The Times*), a 'bad decision' (*Daily Express*), 'high-handed' (*New Statesman*), 'impetuous and arbitrary' (*Nature*, the scientists' journal), 'an elaborate exercise in rationalisation' (*Observer*) and 'the end product of third-class decision-making' (*Sunday Times*). The *Daily Sketch*, a paper which had by no means been wholeheartedly behind the Stansted case, said: 'Nobody wants an airport near their back garden. It's natural that at Stansted, Essex, they are angry about the Government's plan to site London's third airport there. If that were all the great Stansted battle were about, we'd sympathise with the local inhabitants but back the Government. But there's a right and a wrong way to decide when and where to plonk down noisy and land-hungry airports.'

It was the national arguments about the right and the wrong way to decide these things in a democratic, legally-based society that were to dominate the remainder of the Stansted affair. But there was also a new militant mood at the local level. Sir Roger Hawkey voiced it a week later when he told a public meeting attended by some 3,000 people at Great Dunmow: 'we won the public inquiry and the Government's decision was a travesty of justice. We have been cheated. We say to Mr Jay, President of the Board of Trade, that we are prepared to fight this to the last ditch, and then beyond the last ditch.' The local anger comes over clearly in Sir Roger's words but, in fact, there was at Stansted far less evidence of a last-ditch fight in any literal sense of that phrase than there was later at Cublington where the leaders of the Wing Airport Resistance Association were careful not to use words like Sir Roger's. All the same the local campaign was entering a new and angry phase. Four hundred supporters later went with Sir Roger Hawkey and the two local Members of Parliament to see Mr Jay and received little satisfaction from the meeting. 'I went in an angry man,' said Sir Roger, 'and I've come out an angry man.' Ironically, the four hundred had taken the route to London proposed by the Government's experts at the inquiry as the rail link for the third London airport: they arrived at Victoria station three hours later to be greeted on the platform by a man who had made the journey from Sheppey to London in eighty-five minutes. Probably the most important move made locally in the aftermath of the White Paper was to formalise the alliance of local opponents into what was called the Stansted Working Party. It was set up five days after publication of the White Paper. Its members included representatives from the North-West Essex and East Herts Preservation Association, the county councils of Essex and Hertfordshire, the National Farmers' Union, other local authorities and amenity groups and the two local Members of Parliament. They were Peter Kirk, Lord Butler's successor at Saffron Walden, and Stan Newens, the left-wing Labour MP for the Epping constituency which included the New Town of Harlow. The two men were to play an important part in the later parliamentary stage of the affair.

Two signs of local authority activity were soon apparent. In May Essex County Council began an action in the High Court to have the Government's handling of the Chelmsford inquiry result declared *ultra vires*. The county council contended that the Government could not overturn the Inspector's report, as it had, without holding another inquiry to give local objectors a chance to voice their case. A different sort of activity was shown at Harlow. The Labour-controlled town council, which had distributed an anti-airport newspaper to every home, regaled visitors to the Town Show in August with the noise of a VC-10 airliner at 2,000 feet blasting from the loudspeaker system. If the airport plan went ahead, said the council, by 1975 that noise would be heard over Harlow sixty-four times in every hour of daylight peak flying time.

The Government turned a blind eye to these protests and legal initiatives. The President of the Board of Trade, Mr Jay, was as determined as ever that there should be no delay in building the third London airport. When the White Paper was debated in the House of Commons in June, Mr Jay found support in three arguments.[8] The first was that the national necessity mentioned by the local inquiry Inspector, Mr Blake, did exist. Since 1964, the Labour government had been increasingly concerned with balance of payments problems: the Stansted affair reached its height in the months leading up to the 1967 devaluation of the pound. The only way to beat this national economic crisis, it was argued, was to increase Britain's export earnings abroad and here the aviation industry was eager to compete if only it could have its new airport at the only site so far found suitable, Stansted. Mr Jay also maintained that other sites would cause as much noise nuisance.

His most surprising argument was that there was in fact substantial local support for the Stansted plan. During the debate he referred to a petition signed by 5,000 people in support of the plan, to a vote by the Harlow Trades Council and to a letter to *The Times* from four trade unionists in Saffron Walden. The letter welcomed the airport as creating 'new opportunities for us and our children, more variety of jobs, better facilities for shopping, for leisure time and other activities which become available when population in an area expands'. It claimed to represent the views of 'modern ambitious people' and to redress the unbalanced view of local opinion given by the national press. Similar professed aims inspired the group which organised the petition and called themselves the Stansted Area Progress Association. Their nucleus was a group of workers at Stansted Airport, who held their first meeting at the Barley Mow public house in Stansted in February 1965. However, beyond organising the petition and distributing car stickers saying 'YES to Stansted as the Third London Airport', they had little effect. They did not raise funds but relied upon a whip-round to pay their expenses; many of their supporters at Stansted tended to move at short notice to other airports; and they had not been allowed to participate in the Chelmsford inquiry because they were not 'objectors'. As a writer in *New Society* said of them: 'they expect the Government to fight their case for them'. The third group, the representatives of local trade union branches forming the Harlow Trades Council, was more determined but just as ineffective. In view of the part that trade unionists and Trades Councils were later to play in the Roskill affair, it is worth quoting from a Harlow Trades Council press release about what happened to their initiative. They had arranged to debate a pro-airport motion on June 9, the same evening as the Labour-controlled Harlow UDC was holding its debate:

'There were 18 delegates present when the motion . . . was discussed. In the voting, ten were completely in favour, two were reluctant to have the airport but considered the decision was irrevocable and prepared to support the motion, two were against. Four abstained because they had not been mandated (by their union branches): of these, two considered that their members would be in favour, one was entirely against and the other personally against. Those who voted in favour of the airport considered the resultant airport jobs, industrial and warehouse development and clerical needs, would outweigh the disadvantages, which they felt were being overstated and in any case might well be overcome by further developments in noise prevention and vertical take-off. Obviously, planned development of roads and amenities, etc., was essential. The Trades Council closed their meeting early in order to get along to the Town Hall meeting. Listening to the comments at that meeting, the Trades Council secretary felt that the feelings of the Trades Council should be made known. Unfortunately, this could not be made as a statement but had to be put as a resolution on which there was no discussion. The Harlow Trades Council motion was overwhelmingly defeated.'

It was a brave attempt. None the less, clutching at any straw for support, Mr Jay cited this Trades Council motion as evidence of local progressive opinion in favour of the airport. The implication was clear: the objectors were conservatives as well as conservationists. One of Mr Jay's backbench colleagues called them 'rural Luddites'. They could hardly be described as that. On the left wing of the Stansted objectors were not only Mr Newens and the Harlow UDC, but also the Eastern Regional Labour Party and the South Essex and Harlow branches of the Communist Party. Stan Newens' opposition to the Stansted plan was particularly significant and galling to the Government. He was moved by deep conviction: he felt that northern Essex should remain a green recreation ground for his native East End of London and not be allowed to become an urban extension of the East

End. With Peter Kirk, Mr Newens gathered support for an all-party resolution calling upon the Government to hold a fresh inquiry. There were over two hundred signatures, including almost a hundred Labour MPs. When the House of Commons came to the end of its debate, however, the Government imposed a three-line whip. The House divided on a Conservative motion calling for a new inquiry. It was defeated by 303 votes to 238. Ten Labour MPs, including Mr Newens, abstained.

Seven months later the Government did a complete turn-about and ordered the new inquiry, later to be identified as the Roskill Commission. The Labour rank and file were no more consulted over this decision than they had been about any earlier moves on Stansted. Some backbenchers were understandably furious at having been used as 'lobby fodder' for a policy that proved expendable. What then had happened in those seven months? There were three things, the first of which was the fortuitous departure of Douglas Jay from the Board of Trade. He had been dropped by the Prime Minister, Harold Wilson, because of another matter entirely, his outspoken opposition to the Common Market at a time when Labour was preparing Britain's second bid to enter the European Economic Community. Mr Jay's successor was Anthony Crosland, but the high hopes that Stansted people entertained were not to be realised immediately. He delayed putting the special development order for Stansted before Parliament while he 'familiarised himself with all the documents'. In the company of a local MP, he paid a secret visit to the airport site (as a Conservative Minister was later to do at Cublington). The result of all these deliberations was at first only a change in detail to the Stansted plan, a realignment of runways which increased local anger and opposition rather than allayed it. It may, however, have been the best method available to Mr Crosland of reopening an issue that had already been firmly settled in the Cabinet. The second factor was a build-up of pressure from quasi-official sources, which the Government could scarcely ignore. Soon after publication of the White Paper, the South-East Regional Economic Planning Council—a creation of the Labour government—issued a tart statement that they had not been consulted about Stansted at any time. The Essex County Council writ came to court. The Government won the day with a judgement that the Minister had 'unfettered statutory discretion' in planning matters and is under no duty 'to act judicially or quasi-judicially', but the conviction was now spreading fast that this was not the way things should be done. The technical assessor at the Chelmsford inquiry, Mr Brancker, burst into print against the White Paper—first in a letter to *The Times* and then in the *Stansted Black Book*, commissioned by the NWE&EHPA who flew him over from Toronto for the purpose. Then, in March 1968, it was revealed that the august Council on Tribunals had reported to the Lord Chancellor 'as a matter of special importance, that in their opinion the making of a Special Development Order under the Town and Country Planning Act 1962 requires, as a matter of justice, that all whose lives or property will be seriously affected should be given the opportunity of a fair hearing of their objections at a public inquiry'.

By March 1968, however, the battle was over. The third and clinching factor had been the House of Lords. The Conservative-dominated Upper House was already looking for an issue on which they could oppose the Government and force it to reveal its hand over reform of the Upper House. Stansted fitted the bill exactly. As Patrick Gordon Walker relates in his book *The Cabinet*:

'The necessary Order was passed by the House of Commons; but the Cabinet became aware that it might be rejected in the Lords. Rejection of an Order by the Lords was extremely rare and normally would have caused no concern: it could easily be put right by another vote in the Commons. On this occasion public opinion seemed to be stirred to the point that would make difficult a reversal of the Lords' rejection of the Order.'[9]

And so the Labour Cabinet, as Mr Gordon Walker reveals, held its fifth discussion at least of the Stansted issue and this time decided to set up a new commission of inquiry. The commission of inquiry was announced in the Commons by Mr Crosland on February 22. Mr Gordon Walker says that this was seen in the Cabinet as a precursor of the new system of planning inquiries that the Cabinet had already approved. But the Lords also had a hand in the shaping of the Roskill Commission, as it was later to become. Not only Tories, but Labour peers and crossbenchers were demanding that the new inquiry should be headed by a judge of the High Court to ensure a full and fair hearing for all parties involved in the choice of a site for the third London airport. Thus was born the Roskill Commission, with the embryo form of all the elaborate judicial trappings it was later to exhibit.

The form of the Roskill Commission is the subject of the next chapter. But before leaving the Stansted battle one must ask: who did win it, who led the clinching victorious assault? It was won by an alliance and, like all successful alliances, it went on attracting support as it rolled forward until its firepower became irresistible. At first it had been an alliance of determined local objectors. Then it was joined by the two county councils, by other local authorities, by amenity groups, the Press and a nucleus of active MPs. Finally, as the issue became not only a question of the suitability of Stansted or any other site for the third London airport, but also a question of basic democracy and justice, the alliance was joined by almost every voice of importance in the land—except, that is, for the Ministers and the civil servants involved in the discredited decision. So Stansted became a national issue. But that need not belittle the role played from the beginning by the local objectors. As *The Times* said in March 1968: 'the heart of the resistance came from those living in that corner of Essex who did not want their environment changed for the worse. But the strength of their case was that they were able to show that the reasoning supporting the official decision in favour of Stansted was slip-shod and narrowly cast.' If the aviation establishment had done its homework properly, there probably would not have been a Stansted affair or a Roskill Commission. The third London airport might already be in operation—at Stansted.

IV
ROSKILL'S COMMISSION

In 1968, the methods and procedures of the new Roskill Commission seemed very far from 'slip-shod', though even then its terms of reference seemed to some people to be 'narrowly cast'. Without question, it was the most comprehensive, the most ambitious, painstaking, sophisticated and expensive exercise in national planning ever undertaken in Britain. This does not, of course, mean that it could not be wrong and was not, in fact, wrong in its recommendation of Cublington. That was for the Government to decide. The constitutional position was clear from the beginning. The thinking behind the setting up of the airport commission of inquiry was, according to Patrick Gordon Walker, that 'the ultimate decision to be submitted to Parliament must remain with the Cabinet. It is the seat of political authority and a matter of such high importance could not be settled anywhere else. The decision could not be left to the commission of inquiry for that would entail conceding to it the power to determine public expenditure. One of the major factors in the problem was the comparative cost of different alternatives.'[1]

On the other hand, in view of all the circumstances of the commission's establishment there was a distinct impression abroad that a Labour administration at least would be inclined to follow the recommendation of the commission it had set up. As one story had it, a senior Labour minister told Sir Eustace Roskill: 'We will carry out what you recommend.' If there is any truth in that, it is quite clear that Mr Justice Roskill himself never claimed finality or infallibility for his team or its methods, not even for the notorious cost/benefit analysis on which the final recommendation hinged. When critics of the Commission claimed that they could discern a distinct posture of infallibility, even though the actual claim was never pressed, they are talking about something else. The posture, the no-nonsense approach of 'these are the Commission's methods and tools and we must use them to get on with the job', all this was effectively built into the commission of inquiry before Sir Eustace Roskill was appointed as its chairman.

The one thing that all observers of the Commission's work agree upon is that it bore the stamp of its chairman as prominently as the head is stamped on a penny. At the time he was appointed, Sir Eustace Wentworth Roskill was scarcely known outside legal circles. He could be identified as a High Court judge who sat usually in rather intricate commercial cases and as vice-chairman of the Parole Board for England and Wales. He had recently presided over a case that briefly caught the public eye, the dispute concerning more than a million pounds worth of shipping shares between Mr Panaghis Vergottis, a Greek shipowner, on the one hand, and Maria Callas and Aristotle Onassis on the other. Yet neither his face nor his courtroom manner was memorable enough to lend themselves to newspaper profiles or to the sort of caricatures of the Bench that adorn the windows of legal shops near the Law Courts in the Strand. Off the Bench, Mr Justice Roskill was essentially a private personality, a family man, though a member of a distinguished family. His father had been a King's Counsel and his two elder brothers were respectively chairman of the Monopolies Commission and the Official Naval Historian of the Second World War. Eustace Roskill lived with his wife and daughters in Berkshire, within a few miles of his brothers' families, devoting his leisure to gardening, swimming and music.

The first quality that he brought to the task of the Commission on the Third London Airport was a strong desire to dispense justice and to be seen to be doing so, in a way that the Stansted affair had patently failed to do. He was pained when it was suggested early in the Commission's life that it was established only to give a veneer of respectability to yet another decision in favour of Stansted already taken in Whitehall. His desire to do justice to all was later to lead him to bend over backwards to give a fair hearing to the pro-airport minorities at the four sites, out of all proportion to the local support for these groups. On the other hand, Mr Justice Roskill and his fellow Commissioners were uneasy, at times impatient, with the single-minded opposition of the anti-airport groups, especially those in North Buckinghamshire. He had put aside all consideration of 'political' factors and the strength of their opposition (as distinct from the case they were arguing) was a political factor. His other outstanding qualities were a very logical mind and a capacity for great application to the task in hand. At the outset, he must have known far less about techniques like cost/benefit analysis than most of his colleagues. Rather ingenuously perhaps in view of his commercial work, he prefaced an early meeting with the remark that he knew nothing of economics: a colleague replied that this might be attributed to his being a Greek scholar for the Greeks had never learned to count. Mr Justice Roskill worked hard at these unfamiliar disciplines and he expected, even demanded, that not only the Commission's staff but also the contracted experts and witnesses should work equally hard. An expert witness who appeared before the Commission said that at least the chairman appeared to have read every word that was put before him and even checked the witnesses' arithmetic. Mr Justice Roskill set enormous store by justice, logic and hard work and he sincerely believed that his faith in them would be shared by every reasonable man.

One cannot help finding a certain naïvety in this faith in technical and judicial logic to the exclusion of all political considerations. 'We concluded,' he wrote in the final Report, 'that if our final recommendation were to command respect and acceptance it had not only to be as right as the best methods could make it but the reasons leading to our judgement had to be as objective and explicit as possible. Only by this approach would there be any hope of persuading informed opinion that our conclusion should be accepted whatever its degree of popularity.'[2] The chairman himself had made the effort and he expected others to do the same. In the final Report, he has a picturesque, though perhaps pedantic, way of describing the initial difficulties that the Commission faced: 'Theseus required Ariadne's thread to escape from the perils of the labyrinth of the Minotaur which he had overcome. We could recognise the labyrinth and we thought we could recognise the Minotaur. When we started we had neither Theseus to lead nor Ariadne's thread to follow.'[3]

The 'we' mentioned here was the Commission which Sir Eustace Roskill had gathered together in May and June 1968. It was a formidable team and it seemed later to the amenity groups and local authorities which appeared before it that its members fell naturally into two camps. On the one hand, there was a trio of obvious economic expertise to whom the mysteries of cost/benefit analysis seemed to represent a congenial intellectual challenge. They were Alan Walters, the young Cassel Professor of Economics at the London School of Economics, Arthur Knight, then Deputy Chairman and now Finance Director of Courtaulds, the textiles group, and Alfred Goldstein, a partner in the firm of consulting engineers, R. Travers Morgan and

Partners. Of the rest, two members of the Commission appeared to have been chosen for specific skills: David Keith-Lucas, Professor of Aircraft Design at the Cranfield Institute of Technology, and A. J. Hunt, a Principal Planning Inspector of the Ministry of Housing and Local Government. The sixth member of the Commission, Professor Buchanan, was in a class by himself. Arguably Britain's leading planner, author of the influential 'Traffic in Towns' report to the Government in 1963, he was when appointed to the Roskill inquiry Professor of Transport at Imperial College of Science and Technology, London University. All the members of the Commission, except for the chairman and Mr Hunt, carried on their professional duties during the three years of the airport inquiry. The Commission recruited its own research team of some two dozen economists, engineers, operations research assistants, systems analysts, planners and statisticians and appointed Mr F. P. Thompson, a senior economic adviser to the Ministry of Transport, as Director of Research.

The first task of the Commission when it met in private on June 25 1968 was to establish the guidelines it would follow. To some extent they had already been mapped out a month earlier by Mr Crosland when he announced the terms of reference—'to enquire into the timing of the need for a four-runway airport to cater for the growth of traffic at existing airports serving the London area, to consider the various alternative sites and to recommend which site should be selected'.[4] Careful phrases like 'the timing of the need' and 'to cater for the growth of traffic at existing airports' indicate that there may have been some last-minute drafting in the corridors of Westminster as part of the consultation with the Conservative Opposition, which Mr Crosland acknowledged. Another result of this consultation was Mr Crosland's emphasis on the two, 'to some extent conflicting', requirements of the Commission: the need for a speedy decision on one of the most important investment and planning issues of the decade, and the need for adequate representation of all the local interests likely to be affected. As noted in the last chapter, the Conservative peers were already adamant that a High Court judge should head the commission. But it was the Board of Trade which drew up more specific guidelines and these were published in the Hansard record for the day that Mr Crosland made his announcement. The Commission's attention would be drawn, said a memorandum, to the following points: 'general planning issues, including population and employment growth, noise, amenity, and effect on agriculture and existing property; aviation issues, including air traffic control and safety; surface access; defence issues; and cost, including the need for cost/benefit analysis'.[5] Thus some of the most sensitive areas of the Commission's work were mapped out before it was even appointed. There was also a rough schedule of the way it should set about its work. The memorandum decreed (though Mr Crosland told the Commons that he had 'no wish to impose upon it an unduly rigid procedure') that the inquiry should be held in five stages: (1) a general invitation to interested parties to give broad evidence and to suggest possible sites, (2) local hearings at the short-listed sites, (3) detailed investigations and research including visits to airports abroad, (4) further discussions between experts to iron out any difficulties and (5) the final stage of public hearings at which all interested parties could be legally represented if they wished.

As well as these guidelines, inherited from the Board of Trade which saw them as a prototype for future inquiries, the Commission added some of its own:

'(*a*) We must start afresh, unfettered by the past, drawing up our own list of possible sites and devising our own criteria for assessing them;

'(*b*) So far as was possible and was compatible with considerations of national security our work should be done in public;

'(*c*) We should make use of cost/benefit analysis as the best available aid to rational decision-making but the results must be subject to close public scrutiny and discussion before any final recommendation.'[6]

In other words, the Commission did not feel itself bound by the President of the Board of Trade's memorandum, though it could scarcely ignore it without very good reason. What happened over cost/benefit analysis, for example, was that two members of the Board's Economic Services Division, who had carried out the cost/benefit analysis of Stansted, attended the first meeting of the Commission at its new headquarters in Templar House, Holborn, and explained the methods they had used.[7] The Commission and its research team examined these methods and in fact rejected nearly all of them in favour of their own form of cost/benefit analysis. They did not reject the need for cost/benefit analysis as such. There was no alternative, if the Commission was not to fall into the quagmire which had sucked down the Interdepartmental Committee over Stansted.

The next task was selection of the short list. Again there was nothing 'slip-shod and narrowly cast' about the Commission's work, even though it must have seemed at times like the nursery game of 'Eeny, meeny, miney, mo . . . O U T spells out'. The part of the game which goes 'if he hollers let him go' did not of course apply since the Commission's short-list procedure was the one part of the evidence and proceedings that was not published until the appearance of the final Report.

With the short-list selection, there was another broad sweep through the past and potential airfields of southern England—hopefully the last there will be—and this time a rough radius of 80 miles from Central London was covered with no sites being considered within a 30-mile radius of London Airport—Heathrow. The Commission ordered three main sweeps to be made. The first was by the Ministry of Housing and Local Government, which inched its way over the one-inch Ordnance Survey maps of the region with an odd instrument called a 'roamer'. This was a piece of perspex, shaped like a broad-handled spoon, and represented a crude scale version of the airport and its noise imprint. It was moved about, avoiding built-up areas and unsuitable terrain, until it produced thirty-seven sites. Then a researcher from the London School of Economics was set to work looking for physically suitable sites without regard to the noise problem: his search produced forty-eight sites. The Ministry was then asked to relax its criteria a little and have another go: this produced a further twenty-nine sites. Suggestions for sites were also received from outside bodies, like the British Airports Authority, which suggested forty sites. In all, the Roskill 'long list' contained seventy-eight different sites.

Then began the weeding-out process, the game of 'Eeny, meeny, miney, mo'. The long list was narrowed down to a medium list of twenty-nine sites, including Cublington, Nuthampstead in Hertfordshire, Foulness (both as an offshore and a coastal site) and Silverstone on the Northamptonshire–Buckinghamshire border, but not including Thurleigh in Bedfordshire. Various government departments were consulted along with British Rail, the British Airports Authority and the National Air Traffic Control Services. The list was consequently brought down to fifteen, this time including Thurleigh which had gained a high rating on its air traffic suitability. Then began a more detailed assessment of the fifteen sites, with particular emphasis on the assessment of building costs. This was in fact a simpler version of the cost/benefit analysis which was later to be made of the four short-list sites. At this time, too, the members of the Commission went to look for themselves. They descended upon the sites first by road and then by helicopter. The visits went largely unnoticed by the press and public, although some sites, like Silverstone and Thurleigh which had been examined in the latter part of the Stansted affair, were known to be under consideration. It was the cost assessment which was crucial. This brought the list down to five—Cublington, Hockliffe and Thurleigh (both in Bedfordshire), Nuthampstead and Silverstone.

The Commission was now ready for its final juggling act. As there was little to choose between Thurleigh and Silverstone in estimated costs, the latter was dropped. Hockliffe (which is only a few miles from Cublington) gained a low planning rating and this too was dropped. This left Cublington, Nuthampstead and Thurleigh, but no coastal site.

It was at this stage that the Commission came as near as it ever did to breaking up. Foulness had ranked low among the fifteen on cost/benefit assessment and if the Commission was going to stick to its agreed selection procedures then Foulness should not be included in the short list. This was the view of at least two members of the Commission, who reasoned that a site that did not make the grade should not be given an artificial boost in the ratings unless there was a specific political reason for it. There was some debate about whether the Commission should go back to the President of the Board of Trade for further guidance. Alternatively, it was argued, if it excluded Foulness from the short list it would inevitably get the reaction it required, certainly from the press and public and hopefully from the Government too. A third member, however, argued that not only must Foulness be included in the short list but that no inland site should be included; the political cards were too heavily stacked against an inland site, as Stansted had shown. It seemed bound to lead to at least one resignation at an early stage of the Commission's work.

Mr Justice Roskill resolved the dilemma by virtually locking up the Commission, as he might have done an old-fashioned jury, until they reached a unanimous decision. They went to his old college, Exeter College, Oxford, for a weekend during the university's Christmas vacation and there they thrashed it out, with much soul-searching in the intervals between argument. The result was that Foulness as an offshore site was included alongside Cublington, Nuthampstead and Thurleigh. Although it had gained a low rating in the preliminary assessment, it offered a 'different and indeed novel solution to the airport siting problem'.[8] It was a formula which kept the Commission in being but it generated a new set of problems. The 'novel solution' only postponed the political decision between an inland and an offshore site from the short-list stage until after the final recommendation. With the benefit of hindsight, one can see that there ought to have been a reference back to the Government before the short list of sites was drawn up. It would at least have settled the question of whether an inland site would ultimately be politically acceptable, at least to a Labour government.

The short list of four sites was submitted to the President of the Board of Trade in February 1969. The Commission had spent under eight months on this part of its task. Was that too little? It has been argued[9] and is still held in some quarters that this fateful stage of the third London airport search should have received a greater proportion of the Commission's time: it had after all rejected 95 per cent of the possible sites in less than 27 per cent of its time. Two points can be made against this view. It was essential, in view of the sketchy nature of the Interdepartmental Committee's work on Stansted, that a really thorough job should be carried out on the short-listed sites—not only on the technical aspects, but also on consultation with all the affected interests to give them a fair, even a quasi-judicial hearing. Secondly, it can be argued that the results of the short-list selection were confirmed —at least according to the Commission's own criteria—by the subsequent research work and by the public hearings. No subsequent evidence was produced to show that any viable site had been overlooked.

What about Stansted? The press seized upon this spectacular omission from the short list a good month before the list was published. At Stansted, the anti-airport campaigners celebrated with a champagne party—though their victorious emotions were tempered later by the realisation that Nuthampstead just 10 miles away was on the Roskill short list. But why was Stansted omitted? It just did not make the grade, said the Commission. To make sure it had not missed anything, it even went back over the documents of the Stansted affair, including the unpublished papers of the secret inquiry which Douglas Jay had instituted before publication of the 1967 White Paper. Those were the only grounds on which Stansted was not included. The Commission was later to declare that 'those who resoundingly claimed that their past campaigns were responsible for the exclusion of Stansted from the short list could hardly have fallen in greater error'.[10]

Neither then nor later was the Roskill Commission to be swayed by pressures of a political or popular nature. The impression of a popular victory at Stansted persisted however and gave hope to those people who were later to try to exert the pressure that Mr Justice Roskill had turned his back on. With the Roskill Commission they failed; with the Government whose final responsibility the choice of site remained, they succeeded. So at the beginning of March 1969, a year after the end of the Stansted affair, the new battle of Cublington, Nuthampstead, Thurleigh and Foulness began.

V
THE CUBLINGTON COMMUNITY

At the time, few people satisfactorily explained why there was always more opposition to the airport proposals at Cublington than at any other site. Clearly, one reason for it lay with the Roskill Commission. None of the other sites seemed such serious contenders when measured against the Commission's terms of reference and its methodology. Cublington was the nearest site to both London and the Midlands and that seemed to be the key factor. Once the Commission had published the results of its cost/benefit analysis, it followed almost inevitably that it would recommend Cublington to the Government. There was then a feeling that this had to be a fight to the finish, that after the delays and deceptions of the Stansted affair, the Government was bound to endorse the findings of Mr Justice Roskill's open inquiry.

Yet, this cannot be the whole explanation. Even at the earliest stage, before there was any popular understanding of the Commission's methods, there was far more opposition at Cublington than at the other sites. An international airport there had then seemed a frank impossibility: one had only to look at the undulating terrain of the site, part of which had recently been considered for a reservoir, to see how incredible the notion seemed. But far from leading to a complacent expectation that Cublington would drop out of the running, it led to immediate resistance.

The strength of this resistance owed much to the innate resources and character of the people of Cublington. Their inventiveness and their ability to mount a prolonged campaign surprised many observers, some of whom concluded that there must also be unusual financial resources behind it all. It simply will not do, however, to write it off as having been run by the local squirarchy. Cublington was a popular resistance movement, and after Christmas 1970, when the site became the Roskill Commission's firm recommendation to the Government, it became a national resistance movement. But it did not present the same sort of national issue as Stansted had done three years earlier.

No one could accuse the Roskill Commission of failing to do its planning homework or of offending the basic tenets of a democratic society, as the aviation establishment had done over Stansted. The national appeal of Cublington rested solely on what they had to defend and preserve and upon their skill in putting their case across to the nation at large. In many ways, they had more to preserve than the other three sites and more even than Stansted. They could not, it is true, point to the loss of prime farming land: the land in North Bucks, though of good national productiveness, is not in the same class as the broad acres of the Essex wheatland or the flat Hertfordshire country around Nuthampstead. But Cublington, unlike Stansted, would have lost four villages to the airport. They were villages that contained a rich environmental heritage of ancient buildings and charming rural vistas—something they had in common to a certain extent with the site villages at Nuthampstead and Thurleigh—but they also contained communities of outstanding vigour. For many Buckinghamshire people this was their most precious asset.

It was also the most difficult to put across to the outside world. Everyone after all lives in some sort of community to which he feels some degree of attachment. It is difficult to make a complete stranger understand the strength of this attachment. The Sunday tripper from London or Birmingham may stop briefly on his drive through a village and readily agree with arguments for preserving a Norman church or a village green. He can imagine himself living in such an environment. The idea of a village community, on the other hand, if it occurs to him at all, may well be the thing that puts him off the notion of living in the country. The truly rural community—as distinct from the dormitory village or semi-rural housing estate—is alien to the town-dweller in a way that picture-postcard views of village life are not. Yet in an age which has elevated the study of the way people live to a 'social science' and which is fashionably interested in community living, the human disruption at the airport sites should have been a cause of greater concern than the loss of a piece of landscape or of a collection of medieval stones. This was the view of many older villagers who were at times disconcerted by the high-flown environmental and conservationist arguments that became a prominent feature of the Cublington case. They too valued the rural surroundings of their homes but, as is often the case with countrymen, they tended to take these things for granted. What they valued more was the human relationships which surrounded them, the married daughter living down the High Street and the old friend two doors away. The loss of these relationships could have meant the difference between living and a mere shadow of living. They could not have been replaced in a re-housing scheme.

The disruption that the airport would have caused to community life at Cublington was a novel and, in some ways, a controversial feature of the third London airport story. Before the Roskill inquiry, this aspect of major planning had attracted little discussion. To a large extent, it was forced on the Commission by the scale of the airport development and by the fact that selection of an inland site would have entailed an uprooting of whole communities, unknown in Britain in peacetime. The only comparable destruction of whole villages, as at Imber in Wiltshire, had been justified by wartime necessity. Mr Justice Roskill had also been urged in 1968 to include a sociologist among the members of his Commission. Consequently, in the late summer of 1969 he asked three sociologists at Essex University to provide 'information on the disruption of community life on any of the four short-listed sites which would be a consequence of the siting of the third London airport on them'.[1] The rather tortuous circumstances of this request are explained in Chapter VIII but, briefly, the Essex team were to conduct their survey by means of an opinion poll carried out within the 55 NNI noise contours at each of the airport sites. In the event, the results of this survey appeared to play very little part in the Commission's final deliberations. At Cublington, however, they stimulated a lively interest in this aspect of the airport fight that was in some ways at odds with the academic approach of the Essex University team. Some of the newer residents in the area began to examine and tabulate the social and family relationships of the older villagers and came to the conclusion that there was a very special 'texture' of community life at Cublington, endangered by the airport. Few journalists took note of this feature of the local case. One who did was Nicholas Taylor, presenter of the BBC radio programme 'This Island Now', who devoted a special programme to the 'Families of Cublington'. He also wrote a feature about them for the *Sunday Times*, which published a large picture of the inter-related 'root families' of the village of Drayton Parslow.

What then is the 'Cublington community'? There is of course a village of Cublington, a small community of some 190 people, and there are a score of other villages which were similarly threatened by the airport plan. They would have been either destroyed or made virtually uninhabitable by the blanket of aircraft noise. As villages, all of them are clearly defined communities, isolated from other communities by the intervening farmland and each possessing a great many similar institutions, like parish councils, pubs, churches, dramatic societies, darts teams and so on. The village is a clear example of community living. There are other communities at Cublington which transcend the village or parish boundaries. Most farmers and farm workers, no longer the majority of the population, see themselves as a distinct community; so too do many religious groups, like the Methodists, or the people who support the local hunt (the Whaddon Chase), or the footballers, cricket leagues, darts leagues, bellringers or members of the Buckinghamshire Federation of Women's Institutes. All of these are examples of groups that have a wider than parish interest. There was also the much larger community of people who were directly threatened by the airport plan and united in their opposition to it—some 7,500 of them living in a 70-square-mile area of three counties. We can call them a community, as the Roskill Commission did, because they were made into one by the common threat of the airport. Without that threat, they would scarcely have recognised themselves as having much in common. To that extent, the airport plan created a sense of community. It had a similar effect, of course, on the existing communities—the villages and extra-parochial groups—for it made people aware of their affinities with other people whom they may only have been on nodding terms with before. Because of the airport, large public meetings were held, committees were formed to organise 'events' and fund-raising, cadres and cells grew up to press a particular aspect of the campaign. It was, as many people remarked at the time, not unlike Britain in 1940: it took the threat of an outside enemy to make people aware of what they already possessed in common.

This is borne out by the University of Essex study. Its main conclusion was that in terms of the quality of the community life at risk, Cublington had significantly more to lose than either Nuthampstead or Thurleigh; and in terms of the extent of the threatened community, Cublington with its 7,500 people likely to be directly affected had much more to lose than Foulness with its 280 people. (On the other hand, the 'texture' of community life on the isolated Foulness Island was richer than at any inland site.)

The first fact that emerges from the study, carried out in late 1969, is that Cublington is a very stable community. About a third (32 per cent) of its population was born in the villages where they were surveyed. This compares with 20 per cent for the two other inland sites and with fifteen per cent for all English rural districts (as surveyed for the Redcliffe–Maude Commission on the Reorganisation of Local Government). On the other hand, the rather special community at Foulness had a 58 per cent native-born population.

At Cublington, the non-native population showed a greater number of recent arrivals who had moved into the area during the past five years (26 per cent) than in the rest of the country (23 per cent), and a correspondingly smaller proportion who had moved into the area more than five years previously (42 per cent) than in other areas (60 per cent). The Essex sociologists conclude that Cublington in common with Thurleigh and Nuthampstead 'paradoxically . . . can be thought of both as places of relative rapid and recent change and places of relative stability'.[2] The reasons for this, which the study did not go into, are to be found in the rapid improvement of local communications in recent years: the M1 Motorway, which runs near Cublington, was opened in 1959 and the mainline rail link between Leighton Buzzard and London was electrified only in 1966. The 1966 Census showed only a small proportion of the working population commuting to work in the Greater London area: 4 per cent of Wing rural district, which extends from the airport site to the Hertfordshire border and only 2 per cent of all workers in Winslow rural district.

The differences between the sites began to emerge with the Essex team's examination of people's likes and dislikes about their home areas. The highest proportion of Cublington people (58 per cent) liked the area itself, which is marginally *lower* than the other inland sites, and 14 per cent put their greatest value on the people living there, which is marginally *higher* than the other inland sites. As for their dislikes, 71 per cent of the Cublington people found nothing to dislike about the area, which is a significantly higher figure than at any other Roskill site. There were indeed some who disliked the lack of public transport and entertainment, but fewer at Cublington than elsewhere complained of the quietness of the area or its remoteness from their places of work.

In terms of work and of social class, Cublington again stands out from the other sites. It has significantly fewer people engaged in agriculture than the other sites and more engaged in manufacturing, quarrying and transport. As might be expected, it has more working-class residents and fewer middle-class people than the other inland sites.

Using the well-known market research classification of AB, C1, C2 and DE, we get this social spread at Cublington: upper middle and middle class 21 per cent, lower middle 19 per cent, skilled working class 30 per cent and working class and low-income pensioners 29 per cent. At Nuthampstead the spread is 28, 18, 20 and 34; at Thurleigh 33, 18, 22 and 27; at Foulness 12, 6, 18 and 65.

It is when the study turns to the network of kin and friends at the four sites, that Cublington's distinction from the other inland sites becomes most apparent. At Cublington 57 per cent of the people had relatives living in the same parish (as against 41 per cent at Nuthampstead and 40 per cent at Thurleigh); a significant number of them (14 per cent) had five or more families of relatives living in the same parish as themselves and 4 per cent were related to twenty or more families in the same parish. The team did not ask them about relatives living in neighbouring parishes which, in view of the short-distance mobility of Cublington people, was an oversight. The Essex study then sought information about friends and where they lived. At Cublington, 62 per cent of the people surveyed said that all or most of their friends lived in the same parish as themselves and 84 per cent said that all or most of their friends lived in the local area (as against 56 per cent and 79 per cent for Nuthampstead and 53 per cent and 72 per cent for Thurleigh). Forty-two per cent of Cublington people said that all their friends knew each other well (Nuthampstead: 22 per cent; Thurleigh: 28 per cent).

Finally there were questions about participation in local affairs and in the anti-airport campaign. Almost without exception, Cublington people belonged to more trades unions, social and sports clubs, professional groups and other voluntary associations than their cousins at the other three sites. In fact, they had 1·70 memberships per person as against 1·30 at Nuthampstead and 1·17 at Thurleigh. Inevitably, membership of the airport resistance groups was greater at Cublington (22 per cent), than at either Nuthampstead or Thurleigh (17 per cent) or Foulness (10 per cent). This reflected radically different attitudes towards the proposed airport, among both those who belonged to anti-airport groups and those who did not: 80 per cent of Cublington people were 'very much opposed' or 'quite opposed' to the airport, as against 72 per cent at Nuthampstead, 65 per cent at Thurleigh and 68 per cent at Foulness.

The Essex University study marked out Cublington as being different in its community life from the two other inland sites. It was more stable, less agricultural, more working-class, younger, more family-conscious, more tightly knit in its relationships, more active in local affairs and more opposed to the airport. But in other respects, the study was tantalisingly inconclusive. What, for example, was the sociological profile of the airport resistance? In general, it showed that members of the resistance movement were more likely to be older people and more likely to be owner-occupiers than tenants. On the other hand, Thurleigh—the most middle-class of the four sites—had the biggest proportion of owner-occupiers and showed the least opposition to the airport. Again, Foulness was the oldest community both in terms of age and of length of residence, but it registered less opposition than the younger communities at Cublington or Nuthampstead. Indeed, Cublington was a younger community and had fewer owner-occupiers and more council house tenants than Nuthampstead, but was more opposed to the airport. Moreover, there was no internal evidence in the study to explain the richer texture of community life at Cublington than at the other inland sites. It showed all the expected features of a truly rural community—more natives, more contact with friends and relatives, more involvement in local affairs—but yet of course it was not so rural because a smaller proportion of its people were engaged in agriculture. Either agriculture in the Cublington area was organised into larger, capital-intensive units—which was not at all the case—or this was a rural community that had come to rely more on industry than on the land and yet had somehow managed to retain its village-based style of community life.

Up to now, the 'Cublington community' has been discussed in terms of the larger community which was brought into being by the threat of the airport. The 55 NNI noise contour of the proposed airport, within which the Essex team conducted their survey, covered nineteen parishes within a corridor 14 miles long and 5 miles wide. The airport threat apart, there is little that these nineteen parishes have in common. Two major roads, the A5 and the A41, cut across the corridor at either end; the villages at the southern end of the corridor look towards the town of Aylesbury, those at the northern end look towards Bletchley. To capture the true essence of community life on the Cublington airport site, one must look at a smaller area.

At the centre of the corridor are seven parishes which would have been wholly or in part inside or immediately adjoining the airport perimeter. They are Cublington, Dunton, Stewkley and Soulbury within the perimeter and Aston Abbotts, Drayton Parslow and Wing immediately adjoining. Other parishes would perhaps have suffered as severely from aircraft noise, but this group forms a communal entity. Its profile shows all the characteristics found by the Essex team in the larger grouping, in some respects more so. The Essex team compared their data with the returns of the 1966 Census and found, for example, 40 per cent of male workers engaged in manufacturing, 23 per cent in distributive trades, 19 per cent in farming and 9 per cent in transport. The Census returns for the seven villages showed 44 per cent in manufacturing, 20 per cent in farming, 13 per cent in transport and only 15 per cent in distributive trades. The class structure of the seven villages also differed from the larger group, with 46 per cent of male workers being skilled working class as against 38 per cent and correspondingly smaller proportions in other social classes.

The really interesting difference is in the more intense network of family relationships in some of the seven villages. This emerged from evidence submitted to Roskill by Mrs Isobel Smith-Cresswell on behalf of Drayton Parslow Parish Council. A group of villagers had charted the kinship network of the whole village, on the basis of the ties of each household, rather than each individual respondent as in the study of *Disruption of Community Life*. They had found that 51 per cent of the households in the village were 'indigenous' in the sense that one or other of the spouses was born there and that another 21 per cent of households had been resident in the village for more than ten years. Among the non-indigenous households, 7 per cent contained Drayton Parslow people who had 'come back' to the village after periods of family residence elsewhere and a further 10 per cent contained people who had migrated from less than 10 miles away. Indeed, the Drayton Parslow villagers seem to have been fairly constantly on the move. Newly-weds often moved out of the village completely until they could find a house in it, and in the total number of sixty-seven indigenous households, only five individuals had not had at least one change of address within the village. The group then charted the family ties of 202 of the villagers in seventy-nine households (60 per cent of the total number of households in the village). This produced one large network of ties and cross-ties that involved all the 'root families' of Drayton Parslow—Willis, Tattum, Walduck, Bates, Stone and Hounslow—and many of the established surnames of neighbouring villages, like Stewkley, as well. The group admitted that not all these ties were active. 'It would be absurd to suggest that each family had close contact with all relatives, or indeed that they are always prepared to acknowledge them. As most have a substantial number, individual members are in a position to discriminate between them in the way that less stable societies choose their friends. There is an awareness of family links, but sometimes at a fairly casual level. Certainly, never till the Disruption Study questionnaire had they been asked to enumerate them.' Much the same applied to friendships within the village and to involvement in local affairs: it was pointed out that village women would help in almost any village function going, without counting themselves as 'members' of the organisation concerned as the questionnaire had implied. The rich communal life of Drayton Parslow was to a large extent shared by other local villages among the 'inner seven'. It was the sort of communal life that could never have survived the building of the third London airport for it owed much of its strength to the spatial relationships of people. However much care, imagination or money had been spent on rehousing families dispossessed by the airport, this sort of community would have been impossible to re-create.

Yet there is still no explanation of the essential character of these communities and of how they came to survive in an area which is no longer predominantly agricultural. Here it is helpful to concentrate on a particular village and look at the historical depth of its communal roots as well as at their spread. Stewkley, the nearest village to Drayton Parslow, is linked to it by both family ties and in other ways such as sharing a parson. It was the biggest village on the Cublington airport site and indeed the biggest village to be threatened with demolition at any of the four short-listed sites. It is basically, though, a very different village from neighbouring villages. It has long been the boast of Stewkley people that while their neighbours were controlled by squires and landlords, they controlled their own fate. It is a village with a character of its own, as its role in the anti-airport campaign showed. Local people still use an old saying: 'Stewkley! God help us', which is said to have been the answer in bad times if a Stewkley man was asked where he was from (the answer in good times was almost as cantankerous: 'Stewkley! Where d'you think?'). During the anti-airport campaign, the Stewkley style of resistance was sometimes out of step with its neighbours, though in time they came to adopt some of the Stewkley methods, like street demonstrations for example. Similarly, over the course of history, Stewkley as a village without a squire was markedly different in outlook from its neighbouring villages, though again they in time achieved a more independent status and came closer to the Stewkley outlook.

If you visit the village today, you may not be impressed by its beauty. It is not at all compact: it is built along a High Street and indeed claims to be one of the longest villages in the country. Professor Buchanan called it 'long, straggling and in some respects rather ugly', though he added that you could not miss the historical associations behind almost every wall.[3] It is paradoxically a village that is full of history, but also the home of a vigorous, modern community. One could hardly put it in the company of the so-called 'dying villages' of England. By any criteria, Stewkley still has plenty of life in it. If one takes the old but still relevant criteria that Oliver Goldsmith applied to his *Deserted Village*, 'Sweet Auburn', then one looks for the presence of schoolmaster, parson and inn. Stewkley still has its schoolmaster (with a primary school of 124 pupils), its resident Vicar and its four inns—though at one time there were ten pubs and as late as 1955 there were six, which is good going for a village with a strong Methodist tradition. There are in fact two separate Methodist churches, served until the beginning of 1973 by separate circuit Ministers, one of them resident in Stewkley. There are also five grocers' shops, including a Post Office and a branch of the Co-op, a draper's, County Library branch, milkman, baker, electrical engineer, plumber, ladies' and gents' hairdresser, coach proprietor, haulage contractor, six builders and an egg-packing station. Much more was threatened by the bulldozer at Stewkley than its environmental heritage. What the world outside mainly heard about was the threatened Norman parish church—a rare architectural jewel with dog-toothed carvings above its arches, doors and windows but few post-Norman embellishments, not even aisles. But Stewkley church was the symbol of what the villagers were fighting to preserve, not the sum total of their projected loss.

It is clearly a thriving and independently-minded community. But it has been like this for generations and one has to delve into the village's history to understand the roots of this independence. The mention of village history does not mean that the remainder of this chapter will be burdened with the impedimenta of famous names and obscure happenings that usually find their way into village histories. An Edwardian chronicler of Whitchurch, a few miles to the south of Stewkley, began his work with the entry: 'B.C. 55—First Roman invasion of Britain: but it did not reach this district.'[4] Much the same can be said of the impression left by the famous families who lived or owned land in the Cublington area—the de Veres, de la Poles, Hampdens, Churchills and Spencers. Even Mrs Emily Pankhurst, the suffragette, who temporarily hid from her critics in Stewkley, left little impression on the village. What is of relevance to the recent struggle against the very modern enemy of the third London airport, is the way that the villagers coped with former enemies, both inside and outside their community, the way that they formed themselves into hierarchies according to local power or property and the way that historically they resolved their divisions and conflicts.

First, the economy of the village. Until this century Stewkley was, like most English villages, predominantly an agricultural community, but there was also a tradition of local craft and industry going back two or three hundred years. There were for generations two brickworks at Stewkley (in some houses, you can still see bricks that were made before coal was available, where the surface has been glazed by the flaming faggots of furze). Pillow-lace was made here and attempts were made in this century to revive the craft. The women also plaited straw for the buyers from Luton: implements for splitting and trimming the straw were displayed at a village museum in Drayton Parslow to raise funds for the anti-airport campaign. The straw-plait industry existed until a generation or two ago, when the village also boasted wheelwrights, bootmakers, builders of windmills and artesian wells, a blacksmith, a saddler, a tailor, a sign-writer and thatchers, threshers and builders of all descriptions. All of these crafts existed during the lifetime of William G. Capp, one of Stewkley's more famous craftsmen—he was a violin-maker who died in 1972, aged 91.[5] Nowadays, the majority of the village's workers go elsewhere—to the Newton Longville brickworks, the Wolverton railway works, the car factories at Dunstable and Luton and the shops and offices of Aylesbury, Leighton Buzzard, Bletchley, and in a few cases London.

But the natives of Stewkley, though most of them no longer have any contact with farming, still regard the land of the parish as in some sense theirs. They walk or hunt or shoot on it, or merely look at it and pass comment on the state of the crops or livestock. This is natural, for the land is a part of them and of their families. It was the struggles of their ancestors to cultivate and possess the land that moulded the village community of today. And it *was* a struggle. The land is a mixture of heavy clays and clay loams (in the eighteenth century the old name 'Stivclai' was thought to mean 'stiff clay'). It is described locally as good land, but only when it is well looked after and particularly when it is carefully drained. It is an area of smallholdings of 100 acres or less, worked for twelve hours a day and at weekends by husbands and wives. There is a high investment of both capital and labour in this land and with that sort of investment it has a high production of milk, beef, lamb, some pigmeat and poultry. Horses are also bred here and there is some cereal production (about 40 per cent) concentrated on a few farms. Yet for all its disadvantages land here is not cheap and has always been much sought after, even in the days when land could be rented much more cheaply in other parishes than it could be bought in Stewkley. An old Stewkley farmer, Ernie Keen, says that 'in 1910, we paid 30 shillings an acre and the land over there in Wing parish, Lady Wantage's land, were seven and sixpence. That's the rental of the tenant farmers, you see, and they used to find them post rails and gates and all that. Of course we didn't get nothing of that.' The struggle of Stewkley people to own their own land was a long one: there were no dramatic transactions like that at Haddenham, another airport village, where in 1625 the villagers freed themselves of service to their landlords by payment of the considerable sum of £1,532.[6] At Stewkley, it was a continuing process, born partly of historical accident and partly of local determination.

It began, as many things did, in 1066. There was no pre-conquest tradition of peasant freeholding at Stewkley: the only local village to enjoy that was, ironically, Soulbury, which in 1304 passed into the hands of the Lovett family and remained with them until the 1920s. What happened at Stewkley was that soon after the Domesday Book entries, the parish was already split up into four separate manors all farmed by tenants and serfs. This made it easier for the peasant farmers to gain more or less permanent occupation of the land by becoming freeholders or paying fixed fines to the lords of the manor and becoming copyholders. This they did with great frequency. The main legal records for the years 1190–1250 show twenty-five entries relating to property transactions of this type in Stewkley but never more than five for any neighbouring village, including Wing which was slightly larger.[7] Moreover, being situated on a ridge, Stewkley appears to have been little affected by medieval plagues, unlike Cublington which was resited after the Black Death.

The result of this was that by the Tudor period, land and wealth in Stewkley were more evenly distributed than in neighbouring villages like Wing. The *Subsidy Rolls* for 1524 and 1525 show, at the less affluent end of the scale, that fifteen Stewkley householders had property worth

£3, twenty-two worth £2, nine worth £1 and three were assessed on annual wages of £1; at Wing, on the other hand, only four householders had £3 worth of property, fourteen had £2 worth, while forty-one had property worth £1 and seven were assessed on £1 of wages. The rich men of the two villages offer a more dramatic contrast. At Stewkley the richest man was worth £40 (he was a tenant farmer and held most of this land in the hamlet of Littlecote which he enclosed illegally); next to him came a man owning £10 of property. At Wing, by contrast, the wealthiest man held land worth £233, another man held land worth £50, another held land worth £40, two held land worth £30, a sixth held land worth £20, a seventh land worth £12, and then at last we come to a villager worth £10.[8]

By the time one reaches the major period of change in the landholdings of England—the enclosure acts of the eighteenth and early nineteenth centuries—the differences between Stewkley and its neighbouring villages were more pronounced. Whatever the economic effects of parliamentary enclosure on the nation's agriculture, it was a social disaster for the village community: it accentuated class divisions, it put land back in the hands of lay and ecclesiastical landlords who before may have been only titular lords of the manor and it sent many small farmers back to the ranks of the rural labourers. Most of the villages in the Cublington area were enclosed before the turn of the century: Cublington (1770), Soulbury (1772), Wing (1797) and Drayton Parslow (1798). Stewkley withstood this assault longer than most. There was a proposal to enclose part of the parish in 1772, but parliament dismissed the case after a cross-petition from some of the landowners. Another attempt was made in 1803, when the opposition was not so peaceable. Before parliament could accept a petition for enclosure, it was necessary to fix notices of the petition to the doors of the parish church for three consecutive Sundays. In 1803, a solicitor's clerk tried to do this at Stewkley and was violated by a mob of fifty or sixty people, four of whom named Windsor, Belgrove, Foster and Webb were subsequently charged with assault and riot at Buckinghamshire Quarter Sessions and gaoled. A story still current in the village has it that the notice was eventually smuggled into the churchyard by an old woman who hid it beneath her skirts. Whatever the truth of that, Stewkley parish—or 77 per cent of it—was eventually enclosed in 1814.

The Stewkley Enclosure Award is another good indicator of the social structure of the village at the time. Enclosure was intended to concentrate land into larger parcels and fewer hands so that it could be more efficiently farmed than the old strip system of cultivation. The Stewkley Enclosure Award of 1814, however, produced 403 different plots of land, owned by 140 proprietors. The six biggest proprietors held only 31 per cent of the parish after enclosure, or to put it another way 42 per cent of the land actually enclosed. Of these the two biggest landowners were churchmen who were changing their tithes into glebe land—the Bishop of Oxford (385 acres) and the Vicar (288 acres) —both of whom would inevitably let their land to local farmers. The biggest lay proprietor held only 179 acres, or less than 6 per cent of the land enclosed. By contrast, the Wing Enclosure Award of 1797 showed the six biggest proprietors, all laymen, holding 71 per cent of the parish and the biggest of them all, the Earl of Chesterfield, holding over half of the parish. On the other side of Stewkley, at Drayton Parslow, the three biggest proprietors held 81 per cent of the land enclosed.

It would be stretching the point a little far to argue that there is a direct continuity between the successful popular resistance to enclosure at Stewkley and the fight a century and a half later against the airport. The succession of severe agricultural depressions that Britain experienced in the second half of the nineteenth century broke the predominance of the farming interest at Stewkley as in many parts of England. But there are direct links between the two campaigns. For one thing, many of the families who were in the village in 1814 are still there in the 1970s. The stock of the village remains to some extent intact. The other is that the agricultural depressions did not ruin a village like Stewkley, as they did some other villages. Families moved off the land into industry, but remained in close contact with the village and often returned to the land later. Moreover, towards the end of the nineteenth century, farming in this area was given a considerable social boost by a concerted movement towards smallholdings. In 1892, the Manor Farm at Stewkley was split up into smallholdings and offered to labourers from the village under a sort of hire purchase scheme—a down payment of £4 and twenty annual payments of £1 15s (£1·75). Then came Lloyd George's smallholdings act of 1908, under which Bucks County Council acquired a number of farms and split them into smallholdings. Most of these statutory smallholdings are still there in the area north of Aylesbury. There were seventeen of them, totalling 1,136 acres, on the airport site and a further hundred or so council smallholdings under the airport's noise area. The County

Council smallholdings are much sought after. They make an invaluable start in life for the herdsman or labourer who wants to start out on his own. If the airport had been built at Cublington, they could not have been replaced elsewhere in Buckinghamshire.

Another and more radical smallholding scheme took place at Drayton Parslow, where the owners of the largest estates were the Carrington family. At the beginning of this century, the then head of the family—later the Marquess of Lincolnshire—who was in charge of agriculture in the Liberal Cabinet of 1905, handed over much of his land in the parish to small farmers in 40-, 50- and 60-acre lots. He also allowed the parish council to administer a further 200 acres as allotments, so that every household in the parish had some plot of land, either a smallholding or an allotment, on which to keep a pig or two. Lord Carrington also encouraged the formation of a cooperative which provided the villagers with seed potatoes and artificial manure at bulk prices. The cooperative was revived in 1943 as the Village Produce Association and still thrives. Another innovation at Drayton Parslow was the agricultural credit bank, to provide capital for the purchase of livestock or equipment, which was supported as a matter of faith by many local Liberals. The present chairman of Drayton Parslow Parish Council, George Dickens, told me: 'That's what made men independent here, you know. They were all their own gaffers—more so than Stewkley because at that time practically the whole village was smallholdings.'

At Stewkley, as we have seen, the smallholding movement was already in progress. In the 1920s it reached Wing. The last owner of the large estate once held by the Earl of Chesterfield was Lady Wantage, widow of the soldier-politician, Lord Wantage, who had sat in the Commons as Colonel Loyd-Lindsay, VC. She directed in her will that her tenants should be offered their farms at a price proportionate to their rents—an acre rented at 5*s* being offered at £5. The vice-chairman of WARA, Bill Manning, remembers his grandfather standing surety for neighbours who were trying to buy their land under Lady Wantage's will.

All through this period of hardship and change, some families had held on to their land. Here is an example of one family's struggle to do so, told by Ernie Keen:

'I can tell you the history of this farm (at Dunton Road, Stewkley) shortly. My grandfather bought it round about 1870. Now he bought it very dear at a hundred pounds an acre—four thousand pounds, forty acres. Now he had got two thousand pounds that my grandmother had inherited from her father, Mead (the village glazier and plumber), he was a rich man. Well, he bought this and borrowed two thousand pound. He were a cow dealer and went to market and got pneumonia in 1895 and came home and died. At that date this farm was borrowed at eighteen hundred pound and the people who had got the mortgage said they were going to call the money in and have it. And my grandmother walked from Stewkley to Leighton to see a man called Lawyer Willis—another lawyer, he weren't our family lawyer but they'd got to get the money in to work for the people who'd lent it, you see, legally. And he said: Mrs Keen, I will let you have the two thousand pounds but I shall have to have the deeds of the land and he lent the two thousand pounds and paid it all. Well, we'd lost it, but he let us have it for a hundred pounds a year from 1895 to 1910. Well, then that was sold and that made two thousand pound then by public auction. We didn't buy it. Well, then we rented it till 1921 and we bought it again for two thousand five hundred pounds and we're still in it today. So we bought it twice. You see, it's going up and up now, it (the price of land) went the other way then. We bought it and lost it and bought it again.

'Now this other land that's on the Bletchley Road: you can't miss it, it's got "No Never" (an anti-airport sign) on the gate. They call it the Common, that was the village common there's a big gravel pit in it and that was the common gravel pit for the parish. Now, I'll tell you a quick tale about this. My grandfather in 1895 died and they had a forced sale, they had to sell the cattle and everything. One of his pals was a man named Smith, Thomas Smith, a cow dealer from Tinningham, getting near for (Newport) Pagnell. He was so sorry for my people because my grandmother, she'd got a family of eight the oldest was only 20, you see, right down to so high (he gestured below his waist) and he bought one of these cows at the sale and gave it right back to my people, you understand? Well, then, in 1914, we heard that this field was going to be let and my Uncle John who was out of the family farm in housing, he went to Newport Pagnell to see this man to see if he'd let it, and he was most pleased to let him it because he was the man he'd given the cow to, and he took him as a tenant. In 1919, after the first war, he said he should have to sell it and my Uncle John he went down to Newport Pagnell and he bought this field and paid for it for £600. That was the man who'd let him have the cow in 1895 and his nickname, Thomas Smith's nickname was "Gentleman Smith". When the deeds for the field were being examined my uncle was rather alarmed because it showed that out of seventeen acres and a half, ten acres and a half was freehold and the other portion was let on a lease from King James the First in 1603 for six hundred years, but the solicitor advising him said well, you needn't worry about that.

'We were never smallholders, but we have had several holdings. Before we had the Wing Road farm we had a 60-acre holding that is now a county council smallholding. We had Wing Road, this place and the Common and we farmed it altogether. Then one of my uncles died in 1950 and then we divided it between the family, you see.' (Before 1950, Ernie Keen had not been a farmer but a travelling threshing contractor.)

The land and the constant struggle to farm it and own it was one factor in shaping the independently-minded community at Stewkley. Another was religion. Dissent was an early feature of village life, possibly from the time of the Civil War when Cromwell's men are said to have stabled their horses in the parish church. At least, the village was certainly in the parliamentary front line during the early campaigns around Bedford, Banbury and Oxford. With the Stuarts back on the throne, the Established Church set about re-establishing itself in Buckinghamshire, but it clearly had problems at Stewkley. A Visitation by the Archdeacon of Buckingham was made in 1662. At other villages the Archdeacon mentioned individual offenders and specified their penalties. But at Stewkley:

> 'William Wigg William Grace guardiani [churchwardens] presentant the want of a Common Praier booke, booke of homilyes,
> booke of canons, noe surplesse; crave tyme to provide,
> That there are severall anabaptists, quakers and others that
> come not to the publicke assemblyes,
> That there are some children not baptized,
> That the chancell is much out of repaire.'

It would be if Cromwell's horses had slept in it. But the Restoration did not diminish the dissenting tradition of Stewkley. A religious census held in 1676 showed that 14 per cent of Stewkley people were dissenters, a greater proportion than any neighbouring village. A hundred and fifty years later, Stewkley was to prove fertile ground for two further Nonconformist inroads. The Wesleyans came first in the 1830s and the Primitive Methodists in the 1850s. The two congregations still thrive as separate churches at different ends of the village and, though no longer separated doctrinally, it was not until the beginning of 1973 that they came even under the direction of the same Methodist circuit. These churches form a distinct society of their own within the village community. At the High Street Church (ex-Primitives) for example, the name of Faulkner first appears in the records in the 1880s. There are still fourteen Faulkner households—most of them farming families—in regular contact with the church. As the *Methodist Recorder* proudly announced: 'From Soulbury to Stewkley it's almost entirely Methodist farming country. Round every corner one meets a Methodist, nearly always a Faulkner—about forty or fifty of the family in the village.'[9]

In Stewkley today, relations between Methodists and Anglicans are cordial and cooperative. But there are many reports of friction in the past. Methodists are said to have 'baited the parson', whenever they could, over his administration of parish charities. Another source of friction was the Church school, opened in the 1860s. When the school-rate account went into deficit because of the non-payment of some Methodist farmers, the churchwardens decided to hold a sale of farm equipment to bring the account into credit again. At the auction, the Methodist farmers are reported to have crowded the auctioneer in an attempt to stop the sale and relented only when they saw all the bargains going to Anglicans at the back.

It is the divisions and disputes of historical Stewkley that I mention as the last clue to the village's modern character. They rioted over parliamentary enclosure. They also rioted fifty years before when the reform of the calendar robbed them of eleven days: the Vicar rode among the mob on his white horse and read the Riot Act. There was rivalry, not always friendly, between the villagers who lived north of the parish church and those who lived 'down-town' in the area known as 'the City'. There was always opposition to the Hunt in Stewkley, though it was a farmer at Salden, a few miles away, who made the newspaper headlines in the 1920s by warning off the Earl of Rosebery and the Prince of Wales with a shotgun. At Stewkley, there were sportsmen of a rough disposition; there is the tale of the man who for a wager killed a badger with his bare hands, but was scratched and died of lock-jaw. There are stories of recruiting sergeants being routed from the village, of fairground men being thrown into the pond after disputes about money, of the parish council coming to blows. It was obviously a rough community at times—'Stewkley! God help us.' It is not like that now. As the Stewkley Action Committee told the Roskill local inquiry: 'The crime rate in Stewkley is extremely low, as also is the record of delinquency or broken marriages.'[10] But in the conflicts and divisions one can find evidence of community vigour. It is in the 'dying village' or in the big impersonal town that conflict is less important.

This then of all the communities was the one that had the most to lose if the third London airport had come to Cublington. The Stewkley villagers fought the hardest and their schemes for direct action sometimes worried their co-fighters. It was the place from which the majority of plans for militancy, and even violent resistance, originated. In view of the history of Stewkley, of its deep communal roots, these were threats that had to be taken seriously. Mr Hunt, the member of the Roskill Commission who conducted the local hearings, said at Aylesbury that he was in no doubt at all of the feelings of the people of Stewkley. Mr Hunt, perhaps, had other experience of the Stewkley character for he had lived for a period in the 1940s at Dorcas Farm in the neighbouring village of Drayton Parslow.

VI
ORGANISING FOR A FIGHT

Publication of the Roskill short list on Monday, March 3 1969, almost caught the people of North Bucks looking in the wrong direction. Until a few days before, all their fears about the third London airport had been concentrated on Silverstone, the former motor racing circuit just across the Northants border which was known to have been considered by both the post-Stansted inquiry re-examination and the Roskill Commission. A Silverstone Airport Resistance Association (SARA) had been formed and had collected promises of support, mainly in the Buckingham area, although at least one Cublington village, Drayton Parslow, was alive to the Silverstone threat. There had, indeed, been indications that the Roskill Commission was looking at Cublington. It was one of the forty sites suggested to the Commission by the British Airports Authority, as the newspapers had reported, and in late February Arthur Reed, air correspondent of *The Times*, had reported that the Commission was considering Cublington because it gave good access to the capital, the Midlands and the North; was in open country, beyond the air traffic zones; and was 'far enough from any large town to avoid a serious noise problem'. But these heralds of a noisy world were unnoticed. Nor had people noticed the Commission's inspection tour of the Cublington area by car and helicopter. (They were to be much more aware of strangers later: an unfortunate repair man from the Water Board was warned off by a farmer with a shotgun who mistook him for an airport surveyor.) In any case, local farmers were all too occupied with other problems in early 1969, such as the implementation of the Milton Keynes development which was to rob them of 23,000 acres and the Cublington reservoir plans which threatened a few thousand acres more.

The first people to become fully aware of the new local airport site were SARA. The association had been formed five months before on the initiative of a young barrister, Desmond Fennell, and his wife: they had recently run up a hefty mortgage in buying a house at Winslow and were determined not to lose it beneath the shadow of a major airport at Silverstone. One of the first people to join the Fennells in SARA was another barrister, Nevile Wallace who, as head of the planning department of the National Farmers' Union, was privy to the Roskill Commission's work before publication of the short-listed sites. Mr Wallace was, in fact, to be the main provider of information to the new organisation, WARA, during its first months when official details of the airport proposals were not forthcoming. However, on February 27 he told the Fennells that 'Cublington (Wing)' was now the Commission's local choice. On February 28 Alderman Ralph Verney, chairman of the Buckinghamshire Planning Committee, made a speech in which he said there was no foundation to such reports. On the next day, a Saturday, most newspapers firmly predicted that the Roskill Commission would announce its short list on the following Monday and that the chosen sites were Cublington (Wing), Thurleigh, Nuthampstead and Foulness. In effect, the people of North Bucks had two days warning before the official announcement was upon them.

They were, though, a crucial two days. Two farmers who had seen the newspaper reports were Bill Manning of Wing and his neighbour Peter Warren, who only five years before had added 250 acres of the old Wing aerodrome to his own family farm and was still ploughing up bits of the runways. The two men sought out Nevile Wallace, whom they knew through the NFU, and with him discussed the airport threat and their thoughts about opposing it. That same evening, Mr Wallace took Bill Manning along to the winding-up meeting of SARA at Winslow. There they discussed the necessity of giving SARA a 'southwards orientation' and received promises of financial support from some of the people who had already pledged money to the Silverstone cause. For the rest of that weekend Manning was on the phone, calling up all his friends and the contacts he had made from years of helping organise the Buckinghamshire County Show.

So it happened that on the eve of the Roskill announcement the comfortable drawing-room of the Manning farm-house was packed almost to the point of discomfort with concerned and angry people for the first meeting of a new organisation, the Wing Airport Resistance Association. It was officially inaugurated a week later at a meeting in Winslow and unanimously approved by 200 people assembled in the Wing Equestrian Centre on Peter Warren's farm on March 20. It was christened WARA (pronounced 'wearer') because it was the child of SARA; without that connection it would probably have gained a trendier title such as CARE (the Cublington Airport Resistance Executive).

It was realised even at this early stage that an effective resistance to the airport would have to be a two-pronged affair—'like a pitchfork, the countryman's weapon', said someone at Bill Manning's house. One prong of the attack would be the presentation of the WARA case at any hearings held by the Roskill Commission. Desmond Fennell, who was just as anxious to keep the airport from his southern back-door at Cublington as he was to keep it from his front door at Silverstone, seemed best suited to organise this side of the campaign. The other prong would be the organisation of local resistance which was clearly the speciality of Bill Manning. WARA was to maintain this two-pronged approach for the next two and a half years until the battle was won. Its aims, said its constitution, were '(*a*) to ensure that all people who might be affected by the siting of the third London airport (hereinafter called the airport) in the region of Wing be kept fully informed of the implications of this project; (*b*) to co-ordinate and represent public opposition to the construction of the airport, and to ensure that the association's views are presented to any body investigating the siting of the airport, including the Roskill Commission'.

It is interesting to note that, initially, the organisation and co-ordination of local resistance was conceived in terms of either information or support for the presentation of the WARA case at any hearings. The formation of a 'grassroots committee', for example, was planned as an information network, with each village sending one representative to a sort of parliament over which Bill Manning would preside: there had been something similar at Stansted. In fact, the 'parliament' never met and the main task of the grassroots was the collection of signatures for a petition to support WARA's legal case. Little thought was given at this stage to any organised fund raising and the association held itself aloof from any popular demonstrations. Indeed, it tried to discourage them as potentially damaging to the legal arguments. It saw its first function as informative and it performed this from the beginning with vigour. The meeting at Bill Manning's house had already set about producing a pamphlet explaining the implications of the airport plan and calling for resistance. Twenty thousand copies were distributed in the area during the week after the Roskill announcement and 40,000 copies in the weeks following. Ten thousand car stickers were produced and a quarter-page advertisement put in the local newspapers. 'SAY NO TO WING AIRPORT' it proclaimed, and beneath the underbelly of a low-flying jet it advised: 'Don't let it happen. Actively protest. NOW.'

In its first months—indeed in its first year or so—WARA grew very much according to the pattern of other amenity organisations, although its growth was on a scale which was not usual. It still bore a distinct family resemblance to SARA, to the North-West Essex and East Herts Preservation Association at Stansted, on which the Nuthampstead Preservation Association was closely modelled, and to BARA (the Bedfordshire Airport Resistance Association) at Thurleigh. Its leadership followed the classic formula of a president acting as the association's 'Head of State', vice-presidents who stood in for him like a 'Royal Family', a chairman as 'Prime Minister', vice-chairman as 'Chief Whip' and executive committee acting as a sort of 'Cabinet'.[1] WARA's president was David Robarts, the chairman then of the National Provincial Bank, who was a landowner in the northern part of the county and had been president of SARA. The vice-presidents included the chairmen of Bucks and Northants County Councils and six local members of Parliament (four Conservative and two Labour). The chairman of WARA was Desmond Fennell and the vice-chairman, at first, Nevile Wallace, who later had to resign because of his commitments as the NFU's counsel to the other sites. The farmers of Foulness did not particularly relish being represented by the vice-chairman of WARA. He was succeeded by Bill Manning.

For the executive committee, WARA sought out people of administrative experience and prestige and also tried to bring in some new blood, avoiding those who were involved in every other cause in the area. One result of this was that for a considerable period, the committee consisted largely of people unknown to the rank-and-file WARA supporters. It also resulted at first in an apparent social bias towards the wealthier professional members of local society. Among the early members was Andrew Hugh-Smith, a stockbroker and former barrister who later took over from Nevile Wallace the preparation of WARA's information literature and Alec Miscampbell, an Aylesbury solicitor who became known mainly as a public speaker and was thoroughly at home in the later political side of WARA (his brother is Norman Miscampbell, Tory MP for Blackpool). There were two businessmen on the committee, Athel Lonie, marketing director of a Leighton Buzzard firm, and Alan Lovejoy, a public relations consultant who was chairman of another anti-airport group Wings-off-Wing

(WOW) in the villages of Whitchurch, Oving and Weedon. The secretary of WARA was another solicitor, John Pargeter, whose offices in Leighton Buzzard became WARA's headquarters (this office was run by a full-time typist, Mrs Heatherington, who became its lynchpin). The treasurer was Evelyn de Rothschild, the banker, who lives at Wing and the one woman on the committee was Lady (Pamela) Hartwell, wife of the editor-in-chief of the *Daily Telegraph*, who had recently moved to Oving.

These last two members of the executive committee excited some sensational and perhaps jealous comment from outside the area. It was said that Rothschild's Bank was financing WARA and that the editor-in-chief of the *Daily Telegraph* had issued a directive that the paper should carry at least one WARA story every day. Neither was true. The way in which the association raised funds will be discussed later in this chapter, but Lady Hartwell's influence on the communications media appears to have been felt more strongly outside the *Daily Telegraph* than within it. She is an accomplished political hostess and not for nothing a daughter of F. E. Smith. In any case, it would have been truly surprising if the *Daily Telegraph*, in view of its middle-class readership, had not given to WARA the coverage it afforded to other environmental topics. Both Lady Hartwell and Evelyn de Rothschild were introduced to WARA by the Labour MP for North Bucks, Robert Maxwell, whose activities as a vice-president of the association later seemed to fit the role of a rebel prince in the 'Royal Family'. As a result of Mr Maxwell's criticisms of WARA, which will become apparent later, two other members were invited to join the executive committee. They were Dennis Skinner, an insurance agent and trade unionist (he was chairman of the Aylesbury Trades Council) and Walter Randall, a Vauxhall Motors shop steward and leader of the Labour group on Leighton Buzzard town council. Another member, Geoffrey Ginn, the schoolmaster at Stewkley and chairman of the village's anti-airport action committee, joined later as a result of Stewkley criticisms of WARA.

There was initially a serious lack of official information about the airport proposals and one of the first tasks of the members of the executive committee and others was to speak in the villages and towns of the area, night after night, until the people realised what they were up against. It was not until publication of the so-called 'site information', in Cublington's case on April 29, that people had any clear idea of the proposed boundaries of the airport—let alone the noise contours, or the probable area to be covered by service roads, motorways, railways, housing estates, hotels or fuel dumps. Even when the 'site information' was published, much of it was partial and tentative.

The first maps, for example, had shown only part of the village of Stewkley within the airport boundary. David Stubbs, a designer living in the village, took the published specifications for the third London airport, as revealed at the Stansted inquiry, transposed them on to an Ordnance Survey map and decided that not one house in Stewkley would escape demolition. He put up his own sound-contour map of the airport on a large board in his front garden and the Stewkley Action Committee distributed copies of the map to every household in the village. Its broad outline was later confirmed at the Aylesbury hearing. Stewkley, indeed, had a special grievance against the Commission and the media. Some villagers believed that designation of the site as 'Cublington (Wing)' had been a deliberate stratagem to damp down protest from the most populous village to be threatened at any of the four sites. Stewkley rated scarcely a single mention in the press during the first weeks of March, even though a vigorous village action committee was already being formed. A lengthy 'special investigation' of Cublington in the *Daily Express*, for example, gave only two lines to the possible destruction of Stewkley and its Norman church, while it devoted a paragraph to the stately homes in a large area around the airport, and two paragraphs each to interviews with the resident of Ascott House, Evelyn de Rothschild, and to the Vicar of Wing. David Stubbs wrote to the *Beds and Bucks Observer* on March 20: 'I most urgently draw your attention to the plight of STEWKLEY (population eleven hundred) which under the Roskill Committee proposals for the third London Airport at Wing will be entirely obliterated.' Mr Stubbs sent similar letters to sixteen other papers and magazines and to half a dozen well-known broadcasters, and evoked the beginnings of a response. In time, Stewkley was to get all the attention it desired from its organisation of popular demonstrations, but the action committee had to work at it.

Cublington's relations with the press were later to reach a high level of co-operation and understanding—to a large extent through the skill and energy of WARA's part-time press officer, John Flewin, who ran a local freelance agency. But at first, the press, both national and local, painted a confusing picture of the situation at the four Roskill sites. The newspapers had had a good news story placed in their laps by the Commission and they were in danger of smothering it with indiscriminate attention. Their readers were presumed to be anxious to know how the Roskill short list had been received at the grass-roots and the populations of the four sites were certainly not backward in promoting their diverse points of view. But what of the ordinary man honestly in search of information? He would have found it difficult, from scanning the papers, to discover how the airport would affect him, whether it would enhance or ruin his life-style, or whether in the last resort he could do anything to influence the decision.

Another element in the confusion was the widely-aired notion that it was bound to be Foulness in the end. The publicity given during the Stansted affair to the case for an offshore airport and the fact that Foulness had appeared on Roskill's short list encouraged the view that consideration of Cublington could not be wholly serious. People also found comfort in the idea that the Roskill Commission could be railroaded into choosing Foulness by the concerted action of the three inland sites. This was the inspiration of a meeting of Wing Rural District Council two days after the announcement of the short list, when it issued an invitation to the other sites to join forces in pressing for Foulness. The initiative received much publicity, but it stood little chance of success. It could not breach the judicial framework that Mr Justice Roskill had erected for the local hearings, where each site would be considered on its own merits, and only on its own merits. There was to be no repetition of the Stansted inquiry free-for-all. Joint action against an inland site did not get off the ground until the organisation of parliamentary opposition to the Roskill Commission's recommendation of Cublington two years later. There were indeed plenty of red herrings about at the time. The renewal of debate over the third London airport inspired letter writers to take up the cudgels in the press for vertical and short take-off jets, for quieter jets, for airports in the midst of the Surrey Docks, behind St Pancras station and even in Hyde Park and for a return to the days of silent travel aboard Zeppelin-style dirigible airships. The last suggestion gave rise to quite a lengthy correspondence.

It was all so confusing if one had not already made up one's mind. On the eve of the Roskill short-list announcement, the *Sunday Times* had published a summary of the *pros* and *cons* for the four sites by its reporters, Muriel Bowen and Tony Dawe. For Cublington (Wing) it said: '*Pros*: only three miles from main London–Birmingham rail line and twelve miles from M1. New population could be housed in nearby Milton Keynes, Aylesbury and other towns with expansion programmes. *Cons*: hilly area. Extensive bulldozing necessary. Aylesbury and Leighton Buzzard and area of historic monuments affected by noise. Hundreds of houses destroyed to make way for access roads. Complete disruption of air traffic. Right under main London–North airway, and close to holding zone for Heathrow flights.' Was it conceivable that anyone should try to build an airport with all these disadvantages at Cublington? Surely it would not be worth the disruption of all those lives. But when it came to weighing the cost in people against the cost at Foulness in construction money, said the Labour MP for North Bucks, Robert Maxwell, money was bound to win every time. For that reason, and because of its excellent communications with London and the Midlands, he warned that Cublington must be considered an 'odds-on favourite' with Roskill. There was another straw in the wind with publication later on of the Roskill Commission's volume VII of evidence on the 'Proposed Research Methodology'; it outlined the cost/benefit approach and seemed to indicate that even human and community factors could be expressed in financial terms. 'Quite frankly,' said Richard Millard, Clerk to Bucks County Council, 'what is contemplated here sends a cold shudder down one's spine.'

But one family at least was sure. John Taylor, an excavator driver, and his wife had built themselves a bungalow at Burcott on the edge of Wing and decided to assess the airport threat for themselves. They drove to each of the other sites, even going to the edge of Foulness Island to look out across the Maplin Sands, and spent a day at Heathrow to experience the nuisance of living cheek-by-jowl with a major international airport. They came back convinced that Cublington was a strong contender and that an airport there would make local life impossible. The family immediately threw themselves into the WARA campaign—organising meetings, raising funds (even taking an improvised refreshment wagon around the local beauty spots) and spending virtually every leisure hour for the next two years on arranging some event or constructing and painting some sign or poster. Many other local people were similarly convinced of the need for early action. But there was another point of view.

The initial press reports of reactions to Roskill seemed to show a surprisingly large number of people in favour. They proved, however, to be more vocal than numerous. One view was that the third London airport would be no more bothersome and just as good for village trade as the old bomber air station at Wing had been. There was also much talk from this quarter about 'dying villages'. One lady living in Cublington parish, who said that the village was almost dead already, observed that the noise of the airport would not bother her. (Yet no part of Cublington village would have survived.) A more common view came from a man in Wing, who thought 'it is silly to try and stop the inevitable. We should concentrate on making sure the airport causes the least bother possible—and on making a few bob out of it.' This is a view remarkably similar to the feelings of the Harlow Trades Council during the Stansted affair, but at Cublington it was not trade unionists who first organised support for the airport. The Leighton Buzzard Young Conservatives resolved that the airport would be good for local trade but that Foulness was admittedly a better site. It was a topsy-turvy situation, for Leighton Buzzard's Labour MP, Gwilym Roberts, said that the airport would devastate the town, and he called for massive resistance.

A few weeks later the local press reported the formation of the Cublington Airport Supporters' Committee (CASC), with nearly everyone whose names had appeared in the papers as favouring the airport numbered among the committee's officers. CASC made sweeping claims to 'represent the mass of ordinary folk', but at no time did it disclose its list of members nor did it organise a petition in favour of the airport. Indeed, its supporters resisted a postal referendum at Wing. It is difficult to avoid the conclusion that CASC represented not a corpus of opinion in favour of the airport but a few individuals who did not like WARA or the people who ran it. By the time that CASC was formed, it had already reached the peak of its support. It claimed the Duke of Bedford as its patron but was disowned by him and by some of its former supporters during the Aylesbury hearing. At the hearing, however, it managed to steal a good deal of attention, first by complaining to Mr Justice Roskill of intimidation, and secondly by trading on the Commission's desire to give every group a fair hearing.

The growing opposition to the airport is shown by the many attempts to assess public opinion during the first few weeks. A preliminary house-to-house canvass in Wing had shown 502 people against (71 per cent), 112 for and 83 'neutral'. A month later a postal referendum conducted by the Parish Council showed, in the words of the council chairman, 'a majority of 80 per cent against the airport being sited at Wing and there is evidence that opinion has still further hardened against the project'. Attempts at polls were also made by local newspapers and by a Leighton Buzzard urban councillor, but it was not until the late summer that anything like a reliable poll of local opinion was made by the Essex University sociology team. This showed 80 per cent of the people in the wider Cublington area as 'very much opposed' or 'quite opposed' to the airport. No doubt opinion against the airport had hardened, as the Wing chairman suggested, but opposition was all along significantly greater at Cublington than at the other Roskill sites (Nuthampstead 72 per cent; Foulness 68 per cent and Thurleigh 65 per cent).[2]

Official opposition to the airport had already taken a formal shape. Bucks County Council had decided early in the campaign to oppose the airport by all possible means and was soon joined by Bedfordshire, and later by Hertfordshire, Northants and Oxfordshire. The National Farmers' Union formally came out against the airport and in the villages the opposition was becoming organised. Wings-Off-Wing (WOW), which had been formed in Weedon, Oving and Whitchurch along the south-western perimeter of the airport site, saw itself not so much as a WARA village committee but as its local counterpart. It raised £1,000 in cash and promises during its first week of existence: most of the money that WOW passed on to WARA was earmarked for the hiring of a public relations firm.

WARA's own fund-raising effort had been launched at its first formal meeting in Winslow on March 8. In the first week, WARA was able to announce to the press that it had raised £16,500 in cash, cheques and promises to pay for the publicity campaign already underway and for the legal and technical experts to present the WARA case at the Aylesbury hearing. In view of the status of the people concerned, the promises were as good as cash but there were in fact only a few thousand pounds in the bank and the announcement was perhaps ill judged. It gave the impression not only that there was no need for further funds, but that many rich donors had dug deeply into their pockets to launch this organisation. This was to cause many problems inside and outside the ranks of WARA for a very long time to come. It is true that some well-known people of substance were connected with WARA and that its treasurer, Evelyn de Rothschild, was a member of an international banking family. But the mass of the association's funds did not come from sources such as these. It is worth giving some detail of how the funds did come in.

Up to March 20 there had been over 100 donations amounting to some £1,400, of which less than half came in large amounts of £100 or more. On March 20, 48 small donations amounting to £693 were paid into WARA's bank account; on March 24, 84 donations for £507; on March 25, 44 donations for £289; on March 26, 33 donations for £121; on March 27, 32 donations for £166; on March 28, 31 donations for £144; and on March 31, 67 donations for £1,684, of which two were for £750 each and over 50 for amounts of five shillings and ten shillings. WARA was launched on these donations; the proceeds of fund-raising events and activities came in later. A random selection of these events shows £460 from a Stewkley fête, £70 from a Cublington W.I. bingo session, £700 from a sponsored walk at Quainton, £72 from a scrap-metal drive at Great Brickhill, £13. 10*s.* from a coffee evening at Brill and £121. 18*s.* from a 'café continental dance' held by the Haddenham Village Society. Many of these village events, like the sponsored walks, drew supporters from other villages, of course. The list of fund-raising activities went on growing in numbers and in proceeds. A 'Christmas Fayre' at Wing at the end of 1969 brought in almost £2,000, a sponsored walk around the perimeter of the airport site in 1970 netted £3,000, a sale of original drawings given by Fleet Street cartoonists raised £1,850. Of the £57,000 that WARA eventually raised for its fighting fund, about 55 per cent came from marches, rallies, walks—'everything from bingo to beetle drives', said Desmond Fennell. A further £14,000 came from donations of under £100 and some £11,000 came from larger donations—five donations of over £1,000, six of over £500 and twenty-two of over £100.

One of the first calls on these funds was the presentation of the association's case before the Roskill local hearing. The WARA executive committee had decided that its team should be every bit as well-qualified as those representing the local authorities and, perhaps, more able to draw out the broader, national disadvantages of an airport at Cublington. The team, which more than met this brief, consisted of Tom Hancock, consultant planner to Peterborough Development Corporation and Visiting Professor of Urban Planning at Columbia University, New York, Roy Waller, a consulting engineer specialising in environmental problems, and Michael Lamb, an aviation consultant. Desmond Fennell personally had no doubt about who should lead them. Niall MacDermot, QC, formerly Financial Secretary at the Treasury and Minister of State for Housing and Local Government in the Labour government, was already being sought by some of the local authorities for the Aylesbury hearing. He and Desmond Fennell had appeared together for the prosecution in the Great

Train Robbery trial, also held at Aylesbury a few years before. And so Mr MacDermot with Philip Otton became WARA's counsel for all the Roskill proceedings.

WARA was now becoming the focus of immense influence in North Bucks. By June 29—a fortnight before the opening of the local hearing—it had gathered 61,766 authenticated signatures of people who supported its case. It had the additional support of 125 parish councils and 85 other bodies (from women's institutes to the Middle Thames Ramblers). In the space of four months, it had held eighty public meetings in village halls, schools, trade unions halls and clubs of all sorts. Perhaps, in time, the association would exert an important national influence too. It was not surprising that somebody should have made a take-over bid for it. Robert Maxwell, one of WARA's vice-presidents, had been the Labour MP for the airport constituency since 1964. He was already a national personality around whom a certain amount of controversy raged—he had lately made an unsuccessful bid to acquire the *News of the World*, the largest circulation paper in the world—and there was to be further controversy on other matters. But his national image probably counted for less among his constituents than among other sections of the population. They had been served in the past by fairly non-conformist representatives at Westminster: in 1945, their Labour Member was Aidan Crawley, who later became a Conservative MP, and his Conservative successor in North Bucks was Sir Frank Markham, who had been a Labour MP earlier in his career. Mr Maxwell seems to have been accepted for himself by most of the voters of North Bucks and judged on the whole to be a good constituency MP. He had, it is true, been lukewarm towards the anti-Silverstone group, SARA, but at the beginning of the Cublington resistance he played a vigorous role. He was one of the first to warn that Cublington was an 'odds-on favourite' for the third London airport. He had appeared prominently at an early anti-airport meeting in Wing Parish Hall, where he ushered on to the platform Evelyn de Rothschild, who had come down from London in the Maxwell Rolls-Royce. Afterwards, the MP had signed autographs in front of the BBC television cameras, which had filmed the meeting with commentary by David Dimbleby.

Then on Saturday April 12, at another crowded anti-airport meeting, this time in Stewkley, he delivered his bombshell. He said that he had been receiving complaints that WARA seemed to be exclusively a rich man's organisation. 'Unless they can demonstrate that this is false

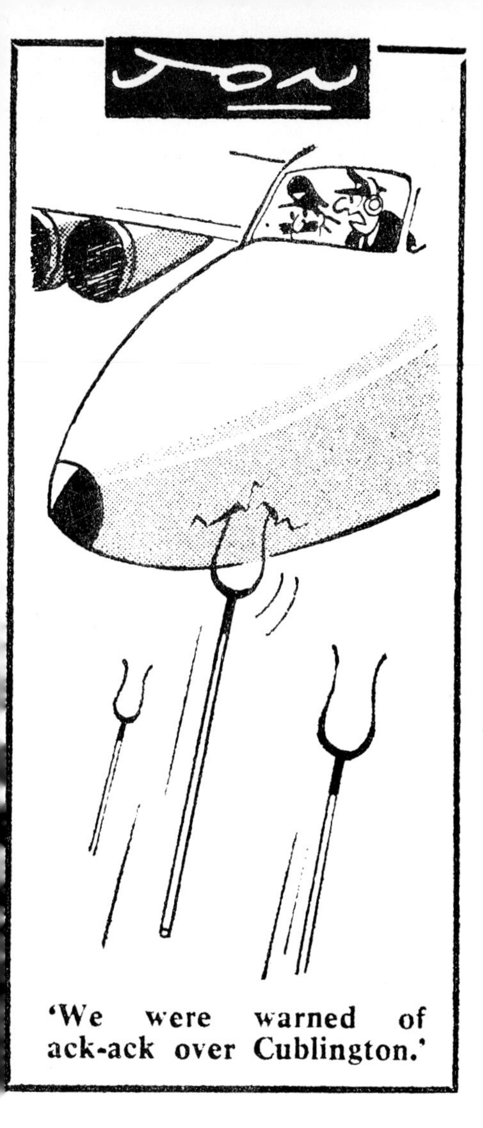

'We were warned of ack-ack over Cublington.'

3

4 *"Well actually a Boeing 707 is as long as a street of houses—this street of houses."*

"I have a feeling this may not be successful . . ."

CUBLINGTON CARTOONS

1 Jon in the *Daily Mail*
2 Graham in *Punch*
3 Colin Wheeler in *Private Eye*
4 Reading in *Punch*

The Roskill Commission (*left to right*): Professor Alan Walters, Professor David Keith-Lucas, Mr A. J. Hunt, Mr D. Caines (secretary),

The WARA executive committee. The chairman, Desmond Fennell, is seen right foreground and the one woman present is Lady Hartwell. (Photo: Press Association)

Above The 'roll-on' of farm vehicles passing through Stewkley, January 1971. (Photo: Associat Newspapers)

Below The Rally. Stephen Hastings, MP, chairman of the parliamentary committee against inland airport, addressing the WARA Rally in the Equestrian Centre at Wing. (Photo: *Lu Evening Post*)

A modern Elijah'. The Rev Hubert Sillitoe igniting the Roskill Report at the WARA rally. Photo: *Sun*)

…al The torchlit procession from Stewkley Church on April 26, 1971 (Photo: *The Times*)

1

2

AIRPORT?
OVER OUR
DEAD BODIES

WARA ART 1 Roskill by Zec 2 Stewkley Man by Colleen Burnett

WARA ART The Norman Church of St Michael and All Angels, Stewkley, threatened wi
destruction by the airport. Drawn by Edward Stamp

Top Another of the churches threatened by the airport, St Martin's Church, Dunton. Drawi
by John Piper

and have representatives of the Labour and Liberal parties on their committee, I intend to dissociate myself from WARA. I will call a special meeting of WARA myself.' A member of the audience told Mr Maxwell to maintain the unity of WARA against the airport and to stop sniping. 'I am not sniping at WARA,' he replied, 'it is they who have been doing it by leaving out all these members.' He asked in particular that the chairmen of Wing and Stewkley parish councils should be more closely associated with WARA village committees. He then accused WARA of having the Rosebery image (Lord Rosebery lives near Wing at Mentmore) and added: 'It is up to them to put this right and I am going to see that they do.'

The incident was given wide coverage in the press and it was of some importance to WARA. In the first place, Mr Maxwell had hit WARA on a weak spot: the leadership was not as representative as it could have been and early steps were taken to redress the imbalance. There was also a lack of communication between the WARA committee and the village activists in the early stages. But the Member of Parliament had used the wrong weapon. WARA was demonstrably not a Tory-controlled organisation. As Desmond Fennell pointed out in a press release, it was non-political and had, in any case, two Labour MPs among its vice-presidents, Mr Maxwell himself and Gwilym Roberts, the Member for South Bedfordshire. Several other people recalled privately that it had been Mr Maxwell in fact who had introduced Lady Hartwell and Evelyn de Rothschild to WARA. As for the Earl of Rosebery, he made his one and only intervention in the affair of the third London airport in a letter to the *Beds and Bucks Observer*: 'There is not one word of truth in what Mr Maxwell said. There is not an inch of land belonging to me which is in any way affected by the proposed Wing airport, nor have I attended the Wing Airport Resistance Association meeting, nor am I in any way connected with it.'

On the Sunday of the following week, Mr Maxwell took his quarrel with the WARA leadership a stage further at a crowded meeting in Desmond Fennell's house at Winslow. He arrived late and, as a WARA minute reports, 'asked if he might interrupt the procedure and stated that he did not consider the association was being run on proper lines, far too much noise was being made too soon, there was complete lack of communication, the association was not properly representative of the area and was being treated as a public relations exercise by a few private individuals. He therefore had no alternative but to resign from the association.' Mr Maxwell's criticisms were taken with the utmost seriousness. He was urged by a number of those present—including Dennis Skinner, David Kessler, publisher of the *Jewish Chronicle*, and Evelyn de Rothschild—not to make his objections a resignation issue. Finally, as the minute goes on to report, 'Mr Maxwell agreed to meet the officers of the association during the following week to discuss his objections. In the meantime, although he would let his resignation stand, he would not announce it publicly.'

Robert Maxwell then left the meeting and those who remained agreed that a letter should be sent to him inviting him to put his objections in writing. A few days later the officers of WARA met Mr Maxwell in London, and his intentions became a little clearer. He told them that they did not have leaders of national standing and reputation on the executive committee and consequently he invited them to resign to make way for such national figures. There was an expectant pause before he added: 'I think I can fairly claim to have a reputation that could sink a battleship.' He then said that, as he was flying to Tokyo on business within a few days, the WARA officers should contact his agent in the constituency by the following Sunday and make to him any proposals they had for reforming the association's leadership: if they failed to do this, they would lose a vice-president. On the appointed day, the WARA officers telephoned the agent who said he had no knowledge of the arrangement.

Mr Maxwell did not in fact resign after this incident, but his business affairs took him increasingly out of the WARA orbit. In July 1969, he was due to address an anti-airport rally in Aylesbury but sent a telegram to say that he could not attend. He was lunching that day with Saul Steinberg, the President of the Leasco Corporation which was later to make a take-over bid for Mr Maxwell's Pergamon Press. Just under a year later, during the general election campaign, the MP was reported by Geoffrey Moorhouse of the *Guardian* as saying one thing about the airport to villagers and quite another to his industrial electors. 'He varies the emphasis of his answers somewhat to suit the company he is keeping. On a high agricultural ridge a gaggle of hunters and shooters are politely, tartly hostile. What does Mr Maxwell think about a proposed Third Airport at Wing? Maxwell emphatically is opposed to such an idea and intends to fight it hard. Earlier in the day, on the (Bletchley) Industrial Estate, a sheet metal worker there had been worried about not getting the airport, and the prospect of 60,000 jobs to build it; he'd have been happier if Maxwell had gone bald-headed for it in the last Parliament, instead of supporting the move to make a new town at Milton Keynes. Maxwell very persuasively suggests that it is better to make sure of the bird almost to hand than of the one that could fly away. Then he adds: "Let's get Milton Keynes first; if we can have the airport as well so much the better." '

That was reported in the *Guardian* on June 15 1970. Three days later, Mr Maxwell lost the North Bucks seat to his Conservative opponent, William Benyon. The Maxwell affair was an odd interlude in the Cublington story, but for WARA it had a special importance. Apart from its immediate effects on the association's organisation and outlook, it was to be WARA's first step in its political education.

One sensitive area that Mr Maxwell had put his finger on was relations between the WARA activists and the elected local authorities. At the Stewkley meeting, where he was perhaps responding to local Labour Party feeling, he had asked that the chairmen of both Stewkley and Wing parish councils should be involved in their village WARA committees. In fact, both chairmen were already members of their village committees. It was true, though, that in both villages the anti-airport activists and the parish council felt unable to join forces: at Stewkley they were rivals in presenting the village's case before Mr Hunt at the Roskill Commission's local hearing. This was not typical of villages in the area. In most of them, the parish council was the focal point of all resistance to the airport and maintained the closest liaison with WARA. At Aston Abbotts, in fact, anti-airport feeling had created a council where none had existed before: a parish meeting in April 1969 decided that the threat to the village called for the creation of a parish council and at the election in the following September the five council seats were contested by twelve candidates, all running on the anti-airport ticket.

It was the relationship between WARA and the Bucks County Council that really raised problems for the association. At first WARA saw its own resistance as adding colour and strength to the County's case. The two champions of the Cublington resistance had started out as if they intended to make the common cause that the NWE&EHPA and Essex County Council had enjoyed over Stansted. But for a time the liaison diminished rather than increased. The question arose, for example, whether WARA would associate itself with the County's planning evidence at the local inquiry, or produce its own planning and other experts. It chose to do the latter. And so the two champions went into battle against the same enemy, each acting independently and each having a full muster of technical witnesses in its support. The reasoning behind this, on the WARA side, was that if the airport were decisively to be beaten, it must be beaten by individual protest, and by arguments on a national scale. There was, in any case, a residue of distrust in the Council's ability to see it through to the finish. The traditional preoccupation of County Hall in Aylesbury with the part of the county to the south of the town, and particularly the County's failure to champion the cause of local residents threatened by the New City of Milton Keynes, had left their mark. Could the County Council ultimately hold out against a development which would seriously affect less than half of its electors, which would help the other half by putting a brake on the expansion of Heathrow, and which would benefit everyone financially by increasing the rateable wealth of Buckinghamshire? Many people thought not. Yet, far from letting down the people of Cublington, the County Council surprised them by the thoroughness and imagination of its campaign. It sent detailed letters of advice on organising their objections to all the smaller councils affected and to all householders on or near the airport site. As a preparation for its case at the local inquiry, it also sent out a questionnaire soliciting objections to the plan from a very wide area. This produced a massive response, as will be seen in the next chapter. It also resulted in some confusion: the County Council had asked people to submit their objections through them, the Roskill Commission was asking for objections to be sent direct to them, and some objectors preferred to send their objections via WARA.

WARA regarded the County as too tied to resistance through institutional channels. It faced a similar criticism itself from the Stewkley Action Committee, which submitted its observations on the reform of WARA at about the same time that Mr Maxwell was making his views known. Whereas the MP maintained that 'too much noise was being made too soon', Stewkley felt that even more noise was needed. Stewkley had discovered that it was threatened with total destruction. It believed that it had been misled by the Roskill Commission and ignored by the media. It was, in any case, a tightly-knit community with an historic tradition of protest and dissent. Little wonder that it should have shown some impatience with the WARA style of resistance, which was felt to be too formal and lacking in scope for individual talents. Stewkley's first venture into the uncharted waters of more vigorous protest was harmless enough. A Mile of Pennies was held in the High Street on Easter Monday and made £63 for WARA.

But its organiser, John Brown, warned the action committee that they had broken a number of traffic laws in the process. The *Daily Telegraph* reported that a human road-block of mini-skirted girls in Easter bonnets had stopped traffic through the village and asked motorists for contributions to the fighting fund. Only one or two refused, it seems. On April 29, the day of publication of the Commission's site information about Cublington, a much bigger event was organised. It was to be a protest march from one end of the village to the other and for this 'spontaneous' protest by Stewkley, press releases were issued in advance. It took place in a carnival atmosphere and brought out 400 people, from other villages as well as from Stewkley. There was a motley collection of 'No Airport' signs and rosettes, as well as the odd pony-rider and placarded family car. The media at last began to notice the village: there were pictures in *The Times* and the *Daily Telegraph* and a good showing on Anglia TV.

Stewkley was still only flexing its muscles. The Action Committee was now almost weekly bombarding WARA with suggestions for improving its communications, with new ideas like the one for a ring of special signs around the airport site perimeter (which was carried out in early 1971), and with demands for information and action. Some activists from other villages and towns had already attached themselves to the Stewkley effort. The action committee, meeting once a week in one of Stewkley's five pubs, had already divided itself into specialised activity cadres. A village newsletter was produced. The committee also distributed 50,000 copies of a pamphlet, giving more technical information about the probable effects of the airport than either WARA or Roskill had yet published. Two young designers, Helen and David Stubbs, formed the St Michael's Church Preservation Committee: in the defence of this rare Norman building, they enlisted signatures from 1,772 visitors to the church, as well as 409 architects, 68 town planners and 45 designers. A door-to-door survey of the village was made and a profile of the community compiled from it for evidence to the Roskill hearing. There was also a survey designed to find out how many villagers would ultimately refuse to leave Stewkley when the bulldozers arrived. It produced a 40 per cent affirmative response. Then there was the formation of a 'Home Defence Group', about which more later. The village was already displaying over a hundred anti-airport signs. The most famous was the representation of an angry farmer standing astride Stewkley church and fending off jet aircraft with a pitchfork. The legend read

'Airport? Over our dead bodies and we mean it.' The slogan caused offence to some Stewkley people. There were all sorts of plans for demonstrations, some of which perhaps fortunately failed to materialise. There was a plan to drive cars and even cattle into Trafalgar Square and cause an anti-airport traffic jam. There was another plan for a group dressed as workmen to descend upon some quiet suburb between Cublington and London and mark out across their unconcerned streets and lawns the 'path of the third London airport motorway'. Stewkley had become a factory of protest ideas and it was turning them out, to the frequent concern of WARA, with great rapidity.

The event that most perplexed the WARA leadership at this time was a protest march through the streets of Aylesbury, planned for the eve of the arrival in the town of Mr Hunt to hold the Roskill local hearing into Cublington objections. The WARA executive was against the march and feared it could detract from the careful preparation and high seriousness of their judicial case. They were partly influenced by a strong (and unjustified) warning that Mr Justice Roskill had given to the Cublington witnesses when he met them at the Commission's London headquarters and produced a letter from a pro-airport resident complaining of 'intimidation'. But with the march the WARA leadership was faced with a virtual *fait accompli*. The Stewkley organisers had already made contact with a number of other villages and received promises of support. In the end, WARA consulted its counsel, Niall MacDermot, and with his approval gave the march its reluctant blessing, provided three conditions were met. The march must be a 'solemn' one, the route must be agreed with the police and Stewkley's estimate of the people who attended must be confirmed by a WARA official. This last condition was consistent with the care with which WARA had checked and double-checked the 62,000 signatures on its own petition.

The march was led by a police sergeant and a loudspeaker van broadcasting 'Land of Hope and Glory'. Behind them marched some 2,000 people, village contingent by village contingent. Undoubtedly there would have been more if WARA had not withheld its blessing until the last moment. Mr Maxwell sent his telegram saying he was unable to attend. There were, however, other famous people in the ranks: the actresses Patricia Neal and Margaret Rawlings, the authors Geoffrey Household and Roald Dahl, and jazz musician Johnny Dankworth and his wife, the singer Cleo Laine. The eighty-year-old Rector of Dunton, the Reverend Hubert Sillitoe, made what the local press called a 'fiery speech' and was followed by Maurice Palmer, aged 16, from Stewkley. It was generally thought to be a satisfactory demo in spite of the arguments over its organisation. A reporter who observed it for *The Times* wrote: 'It was a good-humoured Sunday outing. Many turned out in the deliberate holiday undress of the English middle-classes, all large hats and jeans. One man carried a pitchfork without much conviction.' Cublington people were getting used to that sort of remark from some newspaper reporters.

VII

ENTER THE LAWYERS

The hearing at Aylesbury was the third of the local hearings held at the four short-listed sites. The Foulness hearing had been held in May at Southend-on-Sea, Nuthampstead's in June at Royston, Cublington's in July and the Thurleigh hearing was scheduled for September at Bedford. The staff of the Commission took a short break in August and no doubt they needed it after their experience at Cublington. After an initial burst of enthusiasm for the peculiarities of Foulness—its Brent geese, its cocklebeds—the London newspapers took only a spasmodic interest in these territorial hearings of the Roskill Commission. They were given exhaustive coverage, however, in the local press where this first contact between the Commission and the local populace was seen as the biggest thing that had hit the area since 1940. Yet in a sense nothing had hit them. This was still Roskill's 'phoney war'. Without giving away much information, the Commission was now sending over trial balloons to test the strength of the anti-airport flak that it might get later. They were aptly called 'hearings', for they were not planning inquiries in the sense that the local authorities and many of the individuals present knew them. Despite the status of Mr Hunt as both a member of the Commission and a Principal Planning Inspector of the Ministry of Housing and Local Government (as it then was), there was no one present seeking planning permission for the airport. Neither the Commission itself nor any government agency was putting before the people any firm proposals for the airport development. This was the most striking difference between the Roskill hearings and the local inquiry into the Stansted plan at Chelmsford in 1965. There was just no one present, as there very clearly had been at Chelmsford, for objectors to attack or dispute with. Indeed, the only proposals they could get their teeth into were contained in the so-called 'site information', and that was a very generalised prospectus of what would happen if and when the airport came. In effect, Mr Justice Roskill was saying: 'Let's suppose and assume that my Commission decides that the third London airport should be built at your site. Now let us hear from you your objections, comments and representations on the local consequences.' It was clear that local supporters of the airport were not excluded from this invitation, as they had been by normal planning practice from the Stansted local inquiry. Indeed, the minorities which favoured the airport at Cublington and Thurleigh were proportionately given a far better hearing than the majorities who rejected it. Mr Hunt was fully justified in his assertion that 'nothing had since emerged which would suggest that any interested party was restricted in the full expression of his views whether put orally or in writing'. On the contrary, it appeared that anyone who had something more or less relevant to say, however fallacious or unrepresentative his arguments, was guaranteed a full hearing. The only restriction placed upon evidence at the local hearings was that it must be purely local: there must be no site comparisons at this stage which meant that all mention of Foulness was barred from the inland hearings. This was another difference from the Stansted inquiry. Apart from that restriction, the Commission was prepared to let the local hearings be a sponge, soaking up every opinion and non-opinion before it. Mr Hunt admitted as much: 'the primary purpose of the hearings', he wrote later, 'was to permit as far ranging an expression of opinion as possible rather than to test in every case whether opinion so expressed was well founded'.[1] He put the point more succinctly during a heated moment at the Foulness hearing: 'I am here as the eyes and ears of the Commission, to see the area and to hear what the local people have to say; what they have to say is their business, not mine.' In the circumstances, some of the people in the public seats at the Roskill local hearings wondered how much they mattered in the Commission's scheme for coming to a decision. Were they not, perhaps, designed more as a pressure valve for letting off local steam before it built up in the way it had at Stansted?

The four hearings were very different in local response and in character. This can be shown from the Commission's own figures. Forty-nine parties appeared or were represented at the Foulness hearing, 27 at Nuthampstead, 33 at Thurleigh, but 203 at Cublington. Similarly with written submissions, the Commission received 80 letters at Foulness, 166 at Nuthampstead, 178 at Thurleigh but 725 at Cublington—some 425 of them arising from the Buckinghamshire County Council's initiative. The Nuthampstead and Thurleigh hearings both lasted four days, the Foulness and Cublington hearings lasted eight—but as Mr Hunt told his audience in the Aylesbury Assembly Hall, 'they were not such full days at the other place—you win on points'.

The difference in the character of the hearings was as significant as the variation in response. At the first two hearings, the controversial content was very low-key: indeed at Nuthampstead it was virtually non-existent. Arguments about planning and technical questions dominated. This is probably what the Commission expected, for it had already made clear, when questioned about its decision not to include Stansted on the short list, that it was not to be swayed by popular pressure. The Foulness hearing was unusual in that it presented Mr Hunt with a solid phalanx of institutional *support* for the airport. Essex County Council, Southend Corporation, most of the smaller local authorities, the National Farmers' Union, the Port of London Authority, the Country Landowners' Association and two commercial consortia—the Thames Estuary Development Company (TEDCO) and the Thames Aeroport Group (TAG)—all welcomed an airport on the reclaimed Maplin Sands off Foulness Island. The only argument from this side was that the inevitable increase in the population of south-east Essex should be housed to the south of the River Crouch and as far as possible in existing urban areas. The opposition to the airport from the Foulness Island Residents' Committee and a few neighbouring villages was unco-ordinated and clearly dismayed at their County Council's appearance against them in what they called 'promotion' of the airport. There were more spirited objections from across the estuary in Kent, by the Sheppey Group, than there were from Essex. And in general there was far less argument about the effect on people than about the fate of wildfowl, cockles and the explosive cargo of the wrecked Liberty Ship, *Robert Montgomery*.[2]

At Nuthampstead, the unanimity was all the other way.[3] There was no written evidence and only one witness in person who favoured the airport for its effect upon local employment in this remote corner of north Herts. Indeed, the chairman of the Nuthampstead Preservation Association recounted with some sadness how he had tried in vain to persuade the two or three people he believed to be airport supporters in the village of Barkway to address the annual parish meeting on the subject. Again, the clinching evidence was concerned with planning and technical questions. The land was of high agricultural value; the communications with London and the Midlands were extremely poor. The airport would mean that Cambridge University's £3 million radio telescope, the Decca Company's £250,000 navigator radio station and a number of commercial research laboratories around Cambridge would all have to move. Cambridge University also gave evidence about the effects on its Lord's Bridge telescope at the Thurleigh hearing, which again had its full share of scientific argument.[4] Not only would an airport at Thurleigh outside Bedford create difficulties for the observatory, for the Cranfield College of Aeronautics and for Unilever's agricultural research laboratory, but there had also been a change in the proposed location of the Thurleigh runways since the Roskill short list had been published. This added to the complexities of the arguments at Thurleigh, though that inquiry unlike the first two had its experience of controversial debate. This came from the Thurleigh Emergency Committee for Democratic Action, a body with mainly trade union backing, which certainly kept the anti-airport groups in Bedfordshire on their mettle. At Nuthampstead, though, virtually the only disrespectful remark came from a Queen's Counsel who said that putting an airport there seemed to him like a sick joke. He then recalled that the Roskill Commission did not make jokes.

The Cublington hearing afforded plenty of controversial debate and some hilarity too, though it was far from being a laughing matter for the Commission.[5] The size of the local response was daunting: 811 parties had originally sought leave to take the witness stand, and that would have meant a six-week hearing, even at the brisk pace set by Mr Hunt for his 203 actual witnesses. The response to the Bucks County Council initiative was particularly resented. The 400 or so letters it evoked were privately dismissed by the Commission as emanating from 'Millard's citizens' (after the Clerk to the County Council) and Mr Hunt deliberately put them on one side when he came to assess the balance of Cublington evidence for and against the airport. 'Of all the 300 letters submitted other than those arising from the initiative of the Buckinghamshire County Council,' he wrote, 'only some 2½ per cent were in support of the airport.'[6] Why the discrimination? On his experience of the hearing as a whole, Mr Hunt must have realised that many of 'Millard's citizens' would have written to the Commission even without the encouragement of the County Council. Anyway, 2½ per cent of 300 is only eight pro-airport letters at the most and six of these were from witnesses who appeared at the hearing as the Cublington Airport Supporters' Committee (CASC). Three of these came from the same address in Wing and a fourth from a person related to the family that lived there. But as a percentage of 725, the pro-airport letters amounted to only just over 1 per cent support for the airport, about the same proportion as there was at Thurleigh. At Foulness, 7 per cent of the written evidence supported the airport. It is unfortunate that this perverse computation of the figures by Mr Hunt gave the impression that there was more support for the airport at Cublington than there was at the other inland sites. The Commission's own market research survey showed later that there was considerably less support for the airport at Cublington than at Thurleigh, Nuthampstead or Foulness.

Few of the 725 letters to the Cublington hearing were published in the Roskill evidence. They are all, incidentally, now deposited at the Public Records Office in London and copies of them on microfilm can be seen at the Department of Trade and Industry in Victoria Street. Mr Hunt said that he read them all 'at some risk to my eyesight'. One can quite believe it. They came in all degrees of legibility, from neat typescript and elegant copperplate hand to the laboured jottings of old-age pensioners and manual workers. But they are well worth the strain on the eyes, for they constitute a unique first-hand record of the life of a rural community, and of its hopes and fears. Some of them come straight to the point: 'I am 82 years of age and I was born in North Bucks and I shall not leave here until I am compelled to do so.' Some are carefully argued: 'To take the most optimistic possibility that we find a suitable home elsewhere, we should undoubtedly suffer (*a*) an incalculable social loss, (*b*) a considerable financial loss, (*c*) considerable inconvenience.' Some of them are carefully evocative: 'We chose our home for a life time. We like it. We like the generous rooms and layout. We like the small enclosed garden which is green and mature with flowering trees, flowering shrubs and lilac hedges . . . We are not affluent people. We paid £5,700 for our house and the rest has been achieved by sweat.' Others are quite factual: 'I was born in Stewkley 1898. I have never lived anywhere else or my Father and Mother before me, or my wife. I have lived in this house since my Father died (1941) the first Council house to be built in Stewkley (1928) it has had no other tenant. I have 4 married daughters living in Stewkley or nearby and 4 grandchildren . . . we should think it *very*, *very* hard to be moved.' Many of the letters end with a message that was often heard from the witness box at the Cublington hearing: 'this is my home and I won't leave here except by force.' Even the Clerk to the County Council—the ex-officio 'Clerk of the Peace'—felt moved to remind Mr Hunt that 'this is John Hampden's County and its inhabitants will resist the intrusion of this airport, with all the spirit of that historic figure'.

Although there was little argument at Cublington about airport construction costs or communications or costly scientific establishments (not surprisingly, the Foreign Office offered no evidence about its Diplomatic Wireless Station near the site), the case against the airport was by no means wholly based upon hardships to individuals. Fundamental questions of planning policy were raised and, as Mr Millard said, the building of the airport would entail a 'flagrant violation' of all the principles of planning. Both the County Council and WARA produced expert evidence on this theme. A strategic plan was already being implemented in the Cublington area to house an extra 300,000 people from London and South Bucks in the towns, villages and the New City of Milton Keynes. The area could not accommodate the additional quarter of a million people who would come with the airport without throwing away all pretence at planning strategy. No local settlement could retain its individuality in these circumstances. Milton Keynes would lose all semblance of being a self-contained New City for London's overspill if it were forced to become the airport dormitory. Milton Keynes would inevitably become joined to Aylesbury in a single urban sprawl embracing the airport. The Green Belt would disappear, so would much of the Vale of Aylesbury and many of its villages, so too in all probability would the few green acres that still separate Aylesbury from the advancing conurbation of Greater London. To testify to the amenity value of this countryside not only to its inhabitants but to the nation at large, WARA produced in the witness box two masters of the descriptive word, Sir Arthur Bryant and Sir John Betjeman. Sir John only just made it to Aylesbury: he had left the station at Harrow-on-the-Hill to look at some historic buildings, and had consequently missed his connection and had to complete the journey by taxi. These arguments about planning and amenity were later to be taken up and given extra force by Professor Buchanan as his main argument for opposing the Commission's final choice of Cublington.

All these arguments were delivered with vigour, but it was the throb of popular resistance which showed where the true heart of the Cublington hearing lay. Witness after witness took the stand to say only that he wanted to make it clear he would never leave his home, whatever Roskill and the Government decided. The public gallery booed their Member of Parliament, Mr Maxwell, when he said that the Government's final decision would be accepted; and they cheered the Rector of Dunton, Mr Sillitoe, when he accused the Commission of contemplating an act of sacrilege. As WARA's counsel, Mr Niall MacDermot, QC put it, the Commission's final choice would clearly be influenced by technical and economic studies 'but we all hope that it will never be thought that the emotions aroused in the hearts of people whose lives are going to be affected will never be considered to be a relevant factor'.

WARA had not been happy about the popular emotions expressed at the Aylesbury march, but the main reason for this unease was not lack of sympathy. WARA was concerned that nothing should mar the presentation of its petition of 61,766 authenticated anti-airport signatures. It presented the petition to Mr Hunt in a large wooden drawer, complete with a card index of the signatories showing their geographical distribution; at the same time it handed in documentary evidence of the support for WARA from 111 parish councils and 85 other bodies. Petitions are notoriously unreliable documents and just as easy for the unsympathetic to pick holes in. WARA was aware of both these drawbacks and had recruited a special team of bank clerk volunteers to check on all the signatures. The experience of the anti-airport groups at the other sites is instructive in this respect. The Nuthampstead Preservation Association, having weighed up the merits and demerits of a petition, decided instead to present Mr Hunt with the results of a local opinion survey. At Thurleigh, BARA rather neatly sidestepped the issue when their counsel presented Mr Hunt with a list of 22,000 signatures from 'what are called our supporters'. Thus only WARA with its card-indexed total of almost 62,000 signatures presented the Roskill Commission with a formal petition against the airport. In the event, the WARA petition was not challenged—it would have been very difficult to challenge it—and all the argument about numbers of members, supporters and signatories centred upon the tiny pro-airport group, CASC (the Cublington Airport Supporters' Committee).

The role of CASC at the Cublington hearing is worth some attention, not because they represented any significant body of local opinion in favour of the airport, but because they highlighted a basic weakness in the Roskill local hearings. It was that anyone was allowed to take the floor at the hearing, virtually on his own terms, and especially so if he could claim that he represented an oppressed or ignored minority opinion. CASC had already laid claim to this role before the hearing began, when they wrote to the Commission complaining of abusive telephone calls and threats of physical violence from anti-airport militants. Mr Justice Roskill took these charges seriously enough to deliver a stern warning against intimidation, at a pre-hearing assembly of witnesses at the Commission's headquarters. This was quite unnecessary. No allegation of intimidation was ever proved. But CASC had already won in advance from the Commission a considerable degree of latitude at the Cublington hearing. Six so-called 'members' of the Supporters' Committee appeared at the hearing, but they managed to preserve the right to present themselves as individual witnesses when they chose. Thus, many of the experts called by the County Councils and WARA were subjected to cross-examination by two or three members of CASC in a row, all claiming to speak as individuals. Again, a poultry farmer, who said he was a Vice-President of CASC, gave evidence on one day which he said was on behalf of the Committee (it included incidentally a plan to run the London Underground beneath Cublington and on to Manchester, Coventry and Nottingham). On the following day he declared that his evidence had been given in a personal capacity and that, although he was Vice-President of CASC, he was not in fact an active member of the Committee. All of this took up a considerable amount of time, which seemed to be condoned by Mr Hunt, and it very much annoyed the lawyers and council officials on the other side. The National Farmers' Union extracted a promise from CASC that, as they admitted to no first-hand knowledge of farming, they should not cross-examine the NFU's expert witnesses. Other anti-airport witnesses did not escape so easily. When the time came for the CASC representatives to go into the witness box, the lawyers took their sweet revenge. How many supporters or members does CASC have, they asked. At first, the answers were evasive—'It's another member of the Committee who keeps those figures'—but eventually a figure of 632 emerged. Where is the evidence for this figure, like signatures or membership records?—'They are people who gave their support in the strictest confidence.' Why be so coy about the identity of the support? asked the WARA counsel—'Because there have been numerous not threats but allegations against certain supporters of our Committee, abusive telephone calls, threats also of bad feeling.' What then of the Presidency of CASC, asked the WARA lawyers, are there any letters to confirm your statement that the Duke of Bedford is your President?—'These too are in the strictest confidence.' Are you suggesting, asked the

WARA counsel, that they might threaten the life of His Grace the Duke of Bedford?

There were two sequels to this exchange. The first was the Duke's denial that he had any connection at all with CASC. The other came nine months later, when one of the CASC witnesses told police that petrol bombs had been thrown at his property by 'anti-airport people'. The police investigated and he himself appeared before Buckinghamshire Quarter Sessions and was convicted of malicious damage. None of the allegations of intimidation, in fact, was ever authenticated.

The Commission's permissive attitude towards the Cublington Airport Supporters' Committee was somewhat modified when the time came for the Thurleigh hearing. There the case for the airport was argued by the so-called Thurleigh Action Committee for Democratic Action, a group with claimed the support of fifteen local trade union branches with 8,000 members. But TECDA was represented at the Thurleigh hearing only by its chairman, Mr James Curran, and he alone was allowed to cross-examine anti-airport witnesses under rather stricter conditions than the bevy of CASC representatives had enjoyed.

The odd thing about the permissive attitude of the Roskill Commission towards these minority groups was that it had little bearing at all on the Commission's central purpose of choosing the best site for the third London airport. The Commission was engaged in as objective a search as possible and it had no need to resort to the desperate round-up of every airport supporter that Mr Douglas Jay had resorted to in his parliamentary defence of Stansted. The Roskill Commission was not concerned with choosing a site that would be politically acceptable—in face it chose one that it knew would be politically unacceptable. Why then did it go to such lengths to encourage the Cublington Airport Supporters' Committee? It was a question that frankly puzzled spectators at the hearing in the Aylesbury Assembly Room. The answer is to be found partly in Mr Justice Roskill's own sense of dispensing justice which had been affronted by the (unfounded) allegations of intimidation made before the opening of the Cublington hearing; it was partly in the Commission's desire to leave no stone, however unpromising, unturned, and partly in an antipathy towards the thrusting strength of WARA and indeed of all popular pressure groups.

It was an antipathy that others beside Mr Justice Roskill felt, although the prejudices of one or two London newspaper reporters may have been responsible for this. Whatever the truth of that, the idea is certainly still current that WARA represented only the smart, thrusting, professional middle classes of Cublington and that the mass of ordinary people really wanted the airport and the economic advantages it would bring, but were elbowed out of the limelight by WARA. If there is any evidence to support this view, then it must be very well hidden, for I have not found any trace of it in the research for this book. Quite the contrary. The evidence of the local hearing at Aylesbury, of the written submissions and of the Commission's own public opinion survey carried out at all four sites later in the summer of 1969—all of this evidence shows that the opposition to the airport among all classes was greater at Cublington than anywhere else and the support for the airport correspondingly less. Even the trade unionists, whose brothers at Harlow had voted in favour of Stansted and whose brothers in Bedfordshire were reported to be behind James Curran of TECDA, opposed the airport at Cublington. Shop-stewards' committees were formed in a number of local industries and the Aylesbury Trades Council voted to oppose the airport, though by a narrow majority. The point is that at the local hearings stage of the Roskill inquiry the idea of a community opposing a proposed development with the virtual unanimity of Cublington was still an unfamiliar one. People were ready to assume that there must be two sides to any question. It was not until the end of the following year, after Roskill had recommended Cublington as the airport site, that people became really aware of the massive opposition which had always existed at Cublington towards the airport. This opposition embraced all classes and groups within the community.

VIII

'THE FOUR-SITE SAGA'

An eventful but on the whole frustrating year for the people of Cublington intervened between the end of the local hearing at Aylesbury and the close of Roskill's public hearings in London. It was a year in which the Commission published much bewildering evidence as part of its Stage III Research and Investigations—volume after volume on air traffic, surface access, urbanisation, community disruption, pollution and, most bewildering of all, the results of its cost/benefit analysis of the four sites. It was a year in which this evidence was to be submitted to the grand inquisitorial treatment of thirty-five barristers arguing for seventy-five days in the baroque basement of the Piccadilly Hotel. It was also to be the year of an unexpected, but in the event crucial, general election.

For WARA, though, it was a difficult year. The issues confronting the Roskill Commission had moved away from local affairs, on which many Cublington people felt in any case that they had said all there was to say, to questions of national policy, technology and statistics. The task before WARA was to keep the local opposition on its mettle, to maintain morale and to muster resources for the bigger battles ahead. At the end of August, it launched a new appeal for funds. It had already collected £19,000, of which just under £7,500 had been spent on the legal and technical case at the Aylesbury inquiry (less than had been expected), with £3,300 on publicity and about £1,200 on administration. This left £7,000 in the kitty, but the presentation of the WARA case before the Roskill Commission's long hearings in London could, it was estimated, cost between £20,000 and £30,000. That is roughly what it did cost. So at the beginning of September appeals went out to the 203 village and town committees under the WARA umbrella. The theme was that the response so far had been 'spontaneous and terrific' (secretary John Pargeter's phrase), but since as much again was needed the exercise had to be repeated.

To prime the appeal, WARA showered the area with pamphlets

giving the association's view of the issues already raised by Roskill—pamphlets which bore the distinctive house sign of WARA, the menacing silhouette of a jet with bright red circles beaming from it. The response was encouraging. Villagers again busied themselves at anything that would raise hard cash: fêtes, sponsored walks, piano-bashing contests, coffee evenings, bingo, and as the year wore on carol-singing and Christmas fairs. At Drayton Parslow, the villagers assembled a parish museum of 600 examples of costumes, tools and artefacts. A local historian thought it good enough to rival the County Museum and the chairman of the Aylesbury hearing, Mr Hunt, came down from London to see it. The *Beds and Bucks Observer* caught the mood of Cublington's first autumn under the airport shadow: 'There may not be so many anti-airport posters displayed in windows and on roadsides these days, but the natives are still restless. They must stay that way, for another, longer and more costly Stage V hearing has yet to come and to falter now would prove disastrous.'

Many posters, in fact, had suffered the ravages of an English summer. Repair teams were abroad repainting and replacing the old signs and putting up some not seen before. At Stewkley, possibly the most placarded village in Britain, a curious sign appeared in the Vicarage garden one September evening. It bore inscriptions in Russian, Urdu and Chinese. The Russian text read: 'If it is decided to build the airport here, there will be revolution in Stewkley. We will fight against the government as the Russians did in 1917. It is better to die standing than to live kneeling.' Above these brave words were a few bars of the Red Flag and beneath it the Urdu message: 'The people will never leave here.' The Chinese, which was composed from a Teach-yourself manual, said something to the effect of: 'No surrender. Much resistance.' The author was Arthur Macarthy, who with Martin Evans had organised the Aylesbury march before the local hearing. His purpose was to bring the threatened but obdurate villagers of Stewkley unmistakably to the attention of the press, and he succeeded, but only just. The poster had been erected on a Sunday night and the press had been informed the next morning, but no one had told the Vicar. The Rev Paul Drake, who had only recently come to the village and was not yet privy to the anti-airport stratagems, found the poster on his Sunday night constitutional and took it down. When asked about it the next morning by a caller from a London paper, he reacted with genuine indignation. Later all was explained, the poster went up again and pictures were taken by the local press.

It was an event of marginal significance, but it was a feature of the growing folklore that the airport resistance was now gathering at Cublington and particularly at Stewkley. It also illustrates the popular frustration that was felt at this stage of the Roskill inquiries, the feeling that the experts in London were now making all the running. One of Roskill's specially commissioned expert reports, however, did involve the people directly and it produced unexpectedly heated reactions. The cause was the opinion survey conducted by National Opinion Polls for the team of three Essex University sociologists who were to assess the disruption of community life that the airport would create at the four sites.[1] Local reactions to the survey ran from intense curiosity through puzzlement at the sampling methods and questions (some village women were asked if they belonged to the Townswomen's Guild) to straight anger. In an outcry akin to the reactions that greeted the National Census two years later, eighteen villagers of Aston Abbotts objected to being asked 'how much money they earn per week, where they were last Saturday night, and in which pub they drink'. The villagers told the local newspaper that the survey was 'a diabolical liberty and savours of the Gestapo'.

It was unfortunate that the community disruption study should have got off on this footing, for it was to prove one of the more valuable pieces of research to come out of the Roskill proceedings and, at only one pound, it was by a long way the cheapest. There was, however, an initial failure to communicate the real purpose of the opinion survey and in due course the study itself was heavily criticised both by anti-airport groups and by advocates of the airport, like James Curran's Thurleigh Emergency Committee for Democratic Action. These failings can without exception be traced back to the circumstances under which the study was commissioned. The Commission's first approach to Essex University's sociology department was made at the end of March 1969 and followed by an inconclusive meeting. Essex University appears to have been chosen because it was the source of an earlier suggestion that there should be a sociologist among the members of the Commission. Now the Commission seemed anxious to close a chink in its armour: it was open to the criticism that, the local hearings apart, it had no objective knowledge about the people and the communities the airport might displace. What sort of people were they? Did they go for walks around their countryside, did they have strong local ties or were they mainly commuters who could at a pinch live anywhere? There was a certain amount of sparring about methods of inquiry between the economists on the Commission and the Essex sociologists, and the latter retired feeling there was really little they could contribute. But Mr Justice Roskill was insistent. In May the department was asked if it would be prepared to undertake a survey of the likely sociological effects of the airport and perhaps be able to 'cost the community' that was threatened at each of the four sites. 'Costing community' became a key phrase in view of the Commission's well-known intention to construct a cost/benefit analysis model of the four sites. 'No,' said the sociologists, 'we cannot possibly cost community in those terms.' To this Mr Justice Roskill replied that, if while sitting in the Court of Appeal, he could cost the loss of a leg or an arm or the loss to a woman of her sex appeal, then they could cost community loss. The outcome of this wrangle was that three Essex sociological lecturers agreed to take on the job, on condition that they merely rank-ordered the four sites in terms of the quality of their community life and did not attempt to provide data that could be fed into the cost/benefit analysis model. The main condition of the brief was timing—the study must be finished and presented to the Commission by the end of the year. And so a full-scale sociological survey of four separate communities, embracing 24,000 people—the sort of project on which most university departments would spend at least a year on field work and another year on analysis—was wrapped up in under six months. After the Essex team had undertaken the job, a further month was to be lost while the opinion survey and the data processing were put out to tender. Both went to National Opinion Polls. For the sake of ease in finding a national yardstick for the airport communities, many of the questions in the survey were appropriated from a similar survey undertaken for the Royal Commission on the reorganisation of English local government (the Redcliffe-Maude report). This accounts for such oddities as the question about Townswomen's Guilds. The study itself took the shape of, first, a general profile of each community by age, family structure, housing, length of residence, feelings about the area, income, occupation and class; then an outline of the network of relationships between families, friends and neighbours; and finally an assessment of local participation in public affairs and voluntary societies, including airport resistance groups. This last point distressed the Commission which, as we have seen, was never happy with the political manifestations it produced. The sociologists were forbidden to ask questions about political affiliations or directly about membership of the airport resistance groups, but they got round the latter point by correlating membership of any group concerned with the airport with the respondents' feelings about the airport proposal. By dint of cutting a few statistical corners and by working throughout the night on data collected at Colchester station hot from the London train, the study was completed, printed and formally presented to the Commission by the first week in December. Its broad conclusion, as we have noted in Chapter VI, was that community life at Cublington was in almost every respect more deeply-rooted and more vigorous than at either Nuthampstead or Thurleigh, while Foulness with less than 250 people directly affected compared with 7,500 at Cublington was really a case apart.

Roskill had closed the chink in his armour. From then on the Community Disruption Study seemed to play little part in the Commission's deliberations: no one showed much interest in it at the Stage V hearings and it rated only a single brief mention in the final report. It was published[2] by Her Majesty's Stationery Office on January 29 1970 and immediately eclipsed another report which appeared the following day. This was the long-awaited cost/benefit analysis of the four sites, prepared by the Commission's own Research Team.[3] A great deal of controversy has been attracted to this report. Indeed, it is the target of the main criticism consistently levelled at the Roskill approach, that it tried to summarise all the benefits and 'dis-benefits' of the airport in the everything-has-a-price terms of the market place. This particular criticism is neither fair nor still relevant: Roskill answered it on a number of occasions. Nevertheless, the Commission's use of cost/benefit analysis remains the gravamen of the case against it.

It is worth recalling how it came to use this new and largely unfamiliar tool of public decision making. Cost/benefit analysis was devised at Harvard in the late 1950s as a planning technique for public works and had been used with some success in American studies of water resources and highways and, in Britain, notably in the planning of motorways and the Victoria Line for London's Underground. More recently, the Government has published a controversial cost/benefit study of the Forestry Commission's activities.[4] For Roskill, who was to put it to a far bigger test than any of these, it offered a three-fold attraction. It was the best check available on the public funds to be spent on what was surely the biggest development planned in Britain; it also had the attraction of being a truly objective way of assessing the merits of the various airport sites (as compared with the scarcely objective criteria that had come to light in the Stansted affair); and in any case the Board of Trade had already drawn up an outline cost/benefit analysis for use in the brave new world of large-scale Planning Inquiry Commissions, embodied in the 1968 planning act.[5] The Roskill Commission was to be the testing ground for these procedures which—at the time of writing—have never been used. One can at least agree with Professor Peter Hall of Reading University that 'the Roskill inquiry was a heroic attempt to extend the field of rational, balanced socio-economic inquiry into a very difficult area of decision'.[6]

What is cost/benefit analysis all about? Basically, it is a test of whether a proposed public development, like a motorway or rail-link, will bring the economic and social return that the community can reasonably expect from its investment. It is the equivalent, in the public sector, of the businessman's analysis of profit return on capital. The main use of the technique is to see if a plan is economically and socially viable; its secondary use is to choose between alternative ways of implementing the plan. It does all this by bringing together in one huge equation, one computerised statistical model, not only the tangible items like construction costs, but every conceivable benefit and disbenefit to the community. It should therefore include such things as the effect on other public services, the need for ancillary developments, the savings in freight costs and travelling time, improved or diminished opportunities for new industry and employment, and the liabilities for those people who may have to move or take remedial action, like sound-proofing, because of the development. These factors can properly be measured one against the other only if they are all expressed in comparable terms. If, as Roskill's Research Team did, the analyst chooses to express these disparate factors in the convenient terms of cash, then where does one draw the line?

It was on this point that the Team's *Methodology*, published in June 1969, had resorted to a piece of sophistry that worried many people: 'the absence of market prices does not prevent an assessment of monetary values. If a person is prepared to travel 100 miles in order to visit some feature of archaeological interest, then the value placed on his visit cannot be less than the expenses he is prepared to incur. The attempt to value non-material benefits in monetary terms in no way implies a materialistic view of life. A judge has as a matter of course to make such valuations when assessing damages and this is a socially accepted way of acting. People may assign very high values to cultural or natural phenomena; to observe and record what these values are is not to be confused with pure materialism.'[7] Other authorities on the matter saw it very differently: 'The common value of the £ derives from exchange situations . . . The greater part of the figures used in this type of analysis represent notional values which will never be adequately tested or validated by actual exchanges, and which are highly arbitrary in the sense that a very wide range of values can plausibly be predicated, depending upon innumerable opinions and assumptions. To call these judgements £s is to engage in a confidence trick.'[8] Professor Self, in fact, saw little value at all in the Commission's use of cost/benefit analysis which he described (in Bentham's phrase) as 'nonsense on stilts'. But the use of cost/benefit analysis does not necessarily involve an attempt to translate into monetary terms such subjective valuations as amenity: 'Often it is better not to use such obviously weak evaluations but rather to admit the incommensurable nature of the benefits and discuss the benefits and costs of alternative projects in terms not only of money but of these other physical effects.'[9]

The Roskill Commission, however, refused to admit that some evaluations were incommensurable. When the cost/benefit analysis was published in January 1970, it rejected any idea that there could be two separate balance sheets for the airport—one for the benefits and disbenefits that could be readily expressed in monetary terms and another for more subjective valuations. The line was to be drawn as near to the credible limit as possible. Into what cash terms, then, would the Research Team translate such disparate items as the destruction of a Norman Church, the effect on wild life, the loss of a quiet rural existence or savings in travelling time for both a company director going on business from Manchester to Tokyo and a Bermondsey dustman taking his wife and children on a package holiday to Majorca? This was the second major question of the analysis and all sorts of debatable assumptions were involved in the Research Team's answer.

The Research Team, on the basis of a questionnaire held at the existing London airports at Heathrow and Gatwick, forecast that by the 1990s some 100,000,000 people would be using the third London airport as both passengers and visitors and that many of them would be coming from parts of Britain other than the south-east. Its next assumption was that any savings in the travelling time to and from the airport of these people should be expressed in terms of their incomes. Thus, the travelling time of a businessman, who was assumed to be earning an average of £4,626 a year, was assessed at 100 per cent of his salary or £2·32 (555 old pence) an hour. The income of a leisure passenger (a holidaymaker or sightseer) was assessed at £1,820 a year and his travelling time valued at 25 per cent of this or 23p (55 old pence) an hour. His children's travelling time was assessed at 5½p (13 old pence) an hour. These cash values for travelling time were, of course, highly speculative, particularly so for non-business users of the airport. Since this group of people account for some 75 per cent of the people who pass in and out of a major airport, these speculative sums accumulated into millions of pounds when expressed in terms of the 100,000,000 airport users that the Research Team had forecast for the 1990s. Can one really imagine the average holidaymaker putting a cash valuation on his or his children's savings in travelling time on the airport journey? An hour saved might be of significance, but would five minutes be? Here the Research Team's *Methodology* had produced another curious argument: 'The usefulness of small time-savings depends not on the actual number of minutes saved, but on the opportunity of doing things which might take considerably longer than the original time-saving. For example, 10 minutes saved on the journey to work may make it just possible to take young children to school, something which in itself might take, say, 20 minutes . . . this consideration applies much more obviously in journeys to work, than to holiday journeys.'[10] But most of the airport journeys would be holiday journeys, probably more than 100,000,000 of them a year. Nevertheless, the Research Team costed the five minutes saved by the holidaymaker in a comparable scale to the hour saved by the business-man. Thus, it arrived eventually at a set of large figures called 'passenger–user costs' for each of the four sites. In the case of Foulness out on the coast of Essex, this amounted to nearly half of the total airport costs—£1,041 million out of £2,385 million; for Cublington the passenger–user costs were £887 million, for Thurleigh £889 million and for Nuthampstead £868 million.

There was to be intense debate about these figures at the Stage V Roskill hearings in London. What bothered many people about them was not only the arbitrary assumptions on which they were based—matched in eccentricity only by the decision to cost the loss of Stewkley's Norman church at its fire insurance cover of £51,000—but also their weighting in the overall balance sheet. 'Passenger–user costs' were exceeded only by the 'airspace movement costs', which were shrouded in quite impenetrable mystery. But were they really 'costs'? The Bermondsey dustman on his package holiday would not be paying anyone at all nor receiving payment for five minutes more or less on his journey. On the other hand, the amenity costs, about which much of the third London airport controversy had raged since the days of Stansted, rated comparatively low in the balance sheet. The noise penalty the airport imposed on its new neighbours—even supposing that the law of England were changed to compensate them—was assessed at less than a fortieth of the 'passenger–user costs'. No valuation at all was allowed for the loss to the nation or to posterity of the natural environment, and none for the loss to the people concerned of their community life in the villages and towns. Virtually none of the arguments, produced with such care and at such expense at the local hearings and by airport resistance groups, received a cash valuation in the cost/benefit analysis. Nor was there much joy on the other side for the pro-airport groups: the Research Team gave no valuation to any improved job prospects that the airport might bring to an area. In short, the cost/benefit analysis had confirmed every suspicion that had ever been entertained about it.

But it brought some welcome surprises too. It had shown at least that it was feasible to construct an offshore airport at Foulness—an idea which had begun life as an off-the-cuff suggestion by the Noise Abatement Society—at a cost not much greater than somewhere inland. Construction costs on the reclaimed land at Foulness were assessed as £179 million, at Nuthampstead as £178 million, at Thurleigh as £166 million, while Cublington with all those hollows to be bulldozed in would cost £184 million. Another surprise was the extremely small difference between the total costs of the four sites, when measured against the vast sums involved. Cublington was assessed as the cheapest site at £2,264·6 million. But Thurleigh would have cost only £1·7 million more than Cublington (less than 0·01 per cent), Nuthampstead was £9·3 million higher (0·5 per cent) and Foulness £120·6 million (5 per cent).

The Roskill Commission's attitude to the cost/benefit analysis was one of avowed neutrality. In a foreword, Mr Justice Roskill wrote that the Commission wished 'to emphasise once more that it remains uncommitted to any particular point of view and that it attaches great importance to having these results thoroughly tested at Stage V'. The testing came long before the opening of the Stage V hearings in April, for the cost/benefit analysis was under immediate fire from all sides. The South-East Economic Planning Council attacked the weighting given by the cost/benefit analysis to passenger–user costs 'reaching into a highly speculative future'; such sums it said should not be allowed to swamp the 'immediate and certain burden' the airport would bring to other items in the balance sheet. Moreover, the Research Team had ignored the dynamic economic effects that would spring from an airport at Foulness, combined with a deep-water dock on the reclaimed sands. A similar point was made by Political and Economic Planning, an independent research group. Male unemployment at Foulness, it said, was substantially higher than at the inland sites and there was therefore a greater need of an airport there. The County Councils attacked the analysis for its lack of equity. It seemed to work on the principle of 'unto him that hath (the highly-paid business air traveller) more shall be given, while from him that hath not (the local inhabitant) even his peace and quiet shall be taken away'. Some economists, notably Professor Nathaniel Lichfield and Ezra Mishan, were later to suggest that the costs of the airport as they affected the poor should count for more than costs to the rich. A different point was made by the secretary of the Stewkley church preservation committee, Helen Stubbs, who wrote to *The Times* to protest at the Team's £51,000 valuation of the church. Alderman Ralph Verney put his own figure of £5 million on it: 'I tried to look on it with the eye of someone in the world of art. The church has remarkable examples of Norman arches, which I consider to be each worth the equivalent of a Rubens painting . . . valued at £1 million each.'[11] The critics had shot so many holes in the cost/benefit analysis that it seemed positively daring of Roskill to let it take to the air at the Stage V hearings.

Down at Cublington, however, they were flying other kites. In early April, the *Daily Telegraph* reported that letters calling for 'resistance to the end' were being distributed in the villages by the so-called Home Defence Group and were causing alarm to the members of WARA. A week later, the police were called to the homes of Mr Justice Roskill near Newbury, Berks, and Peter Masefield, chairman of the British Airports Authority, at Reigate, Surrey, to remove signs that had been nailed up during the night. 'No inland airport', they said. In addition 'Not Wing-Cublington' had been daubed in white paint on Mr Masefield's gate. 'This I fear is the life that exists today,' Mr Masefield told the Press. 'People feel that they can damage anyone's property for their own ends.' The incident was roundly condemned by the WARA counsel at the Roskill hearings but the irony of Mr Masefield's words did not go unnoticed.

By this time, the Stage V hearings had opened at the Piccadilly Hotel, London, to much wonderment from the journalists present. 'It really is a quite extraordinary gathering,' wrote Judy Hillman in the *Guardian*, 'that assembled in the nether regions of the Piccadilly Hotel, surrounded by encrusted columns, vast curved mirrors criss-crossed by glazing bars to make them seem more like windows on the world.'

Writing a 'Roskill Commission Notebook' in the *Observer*, Cyril Dunn took the chance of a Sunday newspaperman's greater leisure to describe the 'East-of-Suez ceiling fans', the crash of crockery as the life of the hotel went on in 'other, adjacent bowels of the earth' and the sight of Commissioners and barristers 'seated at parallel tables almost as long as runways and looking rather like a foreign Parliament in session'. The self-consciousness of the gathering came out in a *Sunday Times* photograph of the assembled Roskill Commission seen across the long tables from water carafe level. The paper's reporter, Hugo Young, looked in the other direction at the lawyers who seemed 'happy in their work, carting their briefs about in leather-bound tea-chests'. Another reporter had already recorded the excruciating pun with which counsel for the Commission itself, Mr Arthur Bagnall, QC, had opened the 'four-site saga'. There was an air of concentrated exhilaration about the proceedings. The Commission itself, according to Mr Young, seemed 'preoccupied by the search for a formula of universal validity which will relate every factor to every other and yield a solution which, however unfair or undemocratic or unpopular, will somehow be the absolute and objective truth. They hope to solve the airport problem with an equation as simple and emphatic as Newton's Second Law of Motion.'

This, of course, was parody, but it was parody made possible by the Commission itself. For seventy-five days the 'flower of the English planning bar' (as Professor Self described them) argued about the philosophy and the arithmetic of the cost/benefit analysis while the commissioners sat impassively behind the green baize, occasionally reaching for the water carafe or interrupting to cross-examine. It seemed that every ear in the room was straining to catch the latest nuance to the cost/benefit argument, that every one of the highly-trained minds present was wrestling with the new intellectual problems of this methodology. As Professor Self observed at the time:

'Curiously nobody, save for the Town and Country Planning Association, has chosen to question this basic methodology. Instead, rival economists are being freely employed to tot up the figures in completely different ways. However absurd some of the results, this situation carries certain dangers. First, Roskill may be tempted to argue that, since the basic method has been little questioned, their research team's figures are as good as any and had better be accepted. More broadly, a failure to analyse properly the nature of the Roskill procedures may strengthen the existing tendency to convert genuine political and social issues into bogus technical ones. Finally, the model of decision-making according to which this exercise is being conducted is not, even in principle, an adequately rational one, and its use is blocking more prosaic but more rational methods.'[12]

But not everyone was obsessed by this model of decision-making. In his note of dissent to the final Report, Professor Buchanan freely admitted that, in forming his own view of the problem, he had taken as much notice of a picture hanging in the corridor outside (a reproduction of Rex Whistler's 'Vale of Aylesbury') as of some of the cost/benefit evidence.

There were moments of non-technical drama in the proceedings. The ornithologist, Peter Scott, argued passionately for preservation of the feeding grounds of the Brent Geese on the Maplin Sands. Their worth to the nation was as great as Ely Cathedral, he said; they were as much worth preserving as any 'heap of medieval stones'. Two days earlier, Alderman Ralph Verney, chairman of the Buckinghamshire Planning Committee, had produced a plastic bag containing medieval stones from Stewkley church to show how difficult it would be to move the church and re-erect it elsewhere. But this was an interlude. The proceedings were 'very heavily dominated by the cost/benefit analysis', as Niall MacDermot, QC, counsel for WARA, pointed out, and only at the last stage was the Commission able to address itself to the broader issues of the airport plan.

What many people hoped would be the crucial planning evidence of the inquiry was given by Dr Wilfred Burns, the Government's chief planner. He was director of the high-level planning team charged by the Government with drawing up a strategy for south-east England but, as this team had not completed its work, Dr Burns's evidence was given orally. Planning, it will be recalled, had been the one major element that had been omitted from the cost/benefit analysis, hence the importance of Dr Burns's evidence. It was frankly disappointing. In a nutshell, his evidence was that an airport at Nuthampstead would be the least compatible with the team's planning strategy, but there was little to choose in planning terms between Foulness and the other inland sites. It was the sort of planning assessment that threw the ball back to the Roskill Commission and its statisticians, who received it with the triumphant air of someone who would like to say 'I told you so'. Dr Burns's evidence was, of course, strenuously challenged. Both Essex County Council and the Greater London Council said that they could work up little enthusiasm for Dr Burns's designation of South Essex as a major growth point, unless that growth had a focal point like an airport at Foulness. Industry would not move to South Essex without a major stimulus to the area. Buckinghamshire and other inland counties, with strong support from WARA, argued that Dr Burns was mistaken in his view that there was little to choose on planning grounds between Cublington and Thurleigh. Cublington did not need the stimulus of an airport to grow and to attract industry, whereas Thurleigh clearly did. There was a danger not only of Cublington becoming an overcrowded and overemployed area, but also of its becoming the centre of an urban development stretching right to the edge of Greater London. This was surely no part of Dr Burns's planning strategy for the South-East, said WARA.

On August 10, the shorthand writers put down their pencils and the public hearings were over. Three million words of evidence had been taken from 100 witnesses and 250 documents. Twenty-four organizations had been represented by counsel—seventeen of them favouring Foulness, six favouring an inland site and one, British Rail, remaining neutral. The hearings closed as they had opened, with a joke from counsel for the Commission, Mr Bagnall, QC. There was always the risk, he said, that people would look back on whatever choice was made and see it as 'Roskill's folly'. But, to many people, no folly which the Roskill Commission could cause to be constructed could compare with the intellectual folly of its use of cost/benefit analysis. As Mr MacDermot put it, in his closing speech for WARA: 'If people find that natural beauty and tranquillity and peace of mind, and the relaxed tempo of life which goes with those things, is more important than increased economic prosperity, I don't see how you can quantify that decision in money terms. Such people, I think, would regard the cost/benefit analysis technique as Oscar Wilde regarded the cynic, as the man who knows the price of everything but the value of nothing. This is what I suggest is the limitation of the cost/benefit analysis, that it cannot and does not satisfactorily quantify these amenity considerations.'

So in mid-August, the exotic setting of the Commission's public hearings was returned to the Piccadilly Hotel's guests for their bridge and bingo, and the commissioners retired from public view to compose their report and site recommendation to the Government. But, during the course of the hearings, the complexion of the Government had changed. The Conservatives, who in 1964 had left the Labour government holding the Stansted baby, were now back in office, ready to accept delivery of whatever troublesome infant the Roskill Commission should produce.

IX
HIGHLY RECOMMENDED

The Roskill Commission's recommendation of Cublington as the site for the third London airport was announced by the Secretary for Trade and Industry, Mr John Davies, in the House of Commons on Friday, December 18 1970. It was the last day before Parliament recessed for Christmas. There were echoes of Stansted in the timing. Three years earlier, Douglas Jay had announced the Labour government's decision to press ahead with Stansted on the day before the Whitsun recess, which had been interpreted as an attempt to stave off criticism and had produced uproar. The 1970 announcement was received more calmly, at least by Members of Parliament. Conservative ministers had already made clear their intention of keeping some distance between themselves and the final choice of site until the Roskill recommendation was fully debated. However, the recommendation was made known more than a month in advance of the arguments which supported it, and in Buckinghamshire at least this caused some consternation. Was it another attempt to ward off criticism of an airport decision, by asking people to save their arguments until the full report was published? In fact the reverse proved the case. The recommendation, which was made public before Christmas to prevent distortion of it through leaks in the press, produced such instantly strong reactions that the later revelation of the Commission's full arguments scarcely got the attention they deserved. A great deal of popular activity took place in that intervening month. It will be convenient, however, to take the recommendation and the Report together as if they were one event—which of course they were as far as the Commission was concerned.

The Commission began its recommendation to the Government with virtually the same disclaimer that Julian Amery had used in announcing the Interdepartmental Committee's recommendation of Stansted in 1963: 'There is no ideal site for the third London airport.'[1] If there had been, said Roskill, 'it would have been chosen long ago and there would have been no need for an inquiry on this scale'. The inquiry, as it progressed, had been concerned with two tasks. It was 'an investigation into the largest and most important piece of transport investment that this country, at any rate, had ever seen'; but 'equally true' was its concern with a 'planning decision of crucial importance'. Even when things are equally true, one of them has to be given first mention, and some people saw a particular significance in Roskill's order. It may have been only a minor indication, but then Professor Buchanan was later to say that 'his fellow-commissioners', he thought, 'had looked at the third airport question as an economic problem to which there was an important planning side, whereas his own view was that it was *par excellence* a planning question with an important economic aspect'.[2] If this was a real difference behind the phrasing, then the majority view of the problem went right back to Anthony Crosland's elaboration of the Commission's terms of reference when he told the Commons that 'this is one of the most important investment and planning decisions which the nation must make in the next decade'.[3]

Be that as it may, Professor Buchanan's fellow-commissioners arrived at the heart of the problem, as they saw it, thus:

'A third London airport must be able to succeed as an airport. To this end, it must meet the needs of those whom it is designed to serve. But an airport can succeed as an airport and yet fail in some wider social purpose. This in essence is the case made against an airport at Cublington or Thurleigh. Unfortunately, the converse is not equally true. An airport cannot serve any social purpose unless it first succeeds as an airport. If an airport does not serve its primary purpose satisfactorily, those who are in consequence denied adequate services will go elsewhere and those who look to the airport for some other purpose—for example as an employment base or for alleviating social conditions in east London—will look in vain for that which they seek. Neither in regional planning terms nor in environmental terms can an ailing airport attain the desired result.'[4]

In other words, Roskill's first question was always: can an airport at Foulness succeed as an airport? Professor Buchanan and many other people have said that it could succeed, and should be made to succeed. The Government has said that the airport will go to Foulness and that it will be made to succeed. But even months or years after that decision,

Roskill's question will remain an open one. Foulness could still be the 'white elephant of the century', as Anthony Crosland called it.[5] We shall not know until well into the eighties. It is worth quoting more of this passage from the Roskill Recommendation. It reveals not only the way that the majority of the Commission approached the central problem of inquiry, but also the fundamental dilemma about airport planning which all governments since 1945 have faced—although there is little evidence that they saw the dilemma so clearly before the Stansted affair thrust it upon their attention:

'We do not believe that an airport at Foulness is as certain of success as an airport at either Cublington or Thurleigh. We cannot therefore be confident that the advantage claimed for Foulness will be achieved with that degree of certainty which justifies its recommendation. It is in every sense a relatively expensive project with relatively uncertain prospects of success. When the choice is between two new investments, one economically advantageous and the other less so, we cannot recommend, especially at this point in the nation's history, that the latter should be preferred to the former solely on the ground that the countryside—everywhere within the south-east—and all that goes with it must forever remain immune from the effects of an airport. We view with abhorrence the disturbance of villages, homes and churches, the disruption of communities and the replacement of the quiet of the countryside by the noise of the jet engine. This will occur to a varying extent whichever site is chosen. If we could overcome the dilemma by saying that no third London airport was required, our task would be easy. But a third London airport is required and is required to open its first runway by 1980. The nation's unsatisfactory economic performance in recent years can at least in part be attributed to a national tendency to forgo economic gains and to prefer other goals. Automatic priority should not be given to economic growth. Yet without that growth to which sound investment decisions must make a major contribution, other benefits are apt to disappear like a will-o'-the-wisp. Without such growth the nation cannot afford to spend what it would wish on its environment and on other benefits.'

So the basis of the Roskill recommendation of Cublington was that it made the best economic sense as the site for an airport, even though it might have disadvantages from other points of view. But were any of these other factors important enough to override the economic case and upset the rank-order produced by the cost/benefit analysis? The Ministry of Defence evidence about the Shoeburyness firing range and the Weathersfield air base had been put forward as crucial at the Stansted inquiry. Before the Roskill Commission, the Defence chiefs had admitted that the Shoeburyness range would have to be moved whichever airport site was chosen; now they were more concerned about interference with the Brize Norton air base in Oxfordshire. With commendable scepticism, the Commission wondered whether Brize Norton was not in danger of being made the 'Shoeburyness of Cublington'. Anyway they got from the Vice-Chief of the Air Staff an assurance that there was no overriding Defence ban on the selection of Cublington. This, then, left questions of planning and the environment. It is incorrect and unfair to say, as some critics have done, that the Commission gave less attention to these questions than to economic matters. Most of the debate at the Stage V hearings was concerned with planning priorities and arguments about the environment. There is ample evidence in the Report itself that the majority of the Commission, as well as Buchanan, wrestled hard and long with these questions before concluding that there was no external set of guiding principles to which it could turn for an answer. They did not find it certainly in the evidence of the Government's chief planner, Dr Wilfred Burns, who both in public and private evidence scrupulously refused to give the Commission a positive lead in planning matters. On environmental questions, the Commission had heard what it regarded as such subjective and partisan arguments that it was thrown back wholly on its own resources. As *The Times* remarked: 'The Commission is far from insensitive to the environmental and personal implications of the alternatives, although in contrast to the certitude being expressed on all sides, it confesses to a certain agnosticism.'[6] *The Times*, incidentally, was one of the few papers to allow itself to be led step by step along the path of Roskill's logic—to such an extent that it underwent a conversion from the title of its December leading article, that Cublington was 'The Wrong Place to Put It', to its affirmation in January that 'The Balance of Argument Goes against Cublington.'

Roskill's 'agnosticism' was certainly expressed with sensitivity:

'Each generation is accountable to succeeding generations for the choices which it has made. As William Morris said in 1877: "It has been most truly said that these old buildings do not belong to us only: that they belonged to our forefathers and they will belong to our descendants unless we play false. They are not in any sense our property to do as we like with them. We are only trustees for those that come after us." We cannot know whether our descendants would commend our judgement the more for preserving Stewkley Church or for preserving the Essex coastline and for preventing the extinction of the dark bellied Brent Goose.'[7]

Perhaps a hint of partiality had crept in here for it was the Essex coastline *and* the 'dark bellied Brent Goose', but on the other hand only Stewkley Church, not the Vale of Aylesbury *and* the unspoilt Norman church at Stewkley. However, the point is that Roskill and his colleagues threaded their way through all these 'certitudes' on the value of the countryside, of ancient buildings, rural life, wild life and so on, but felt unable even to attempt to forge from them a scale of priorities of their own:

'As with much else in this inquiry there is no single right answer however much each individual may believe that there is. For us to claim to judge absolutely between these views is to claim gifts of wisdom and prophecy which no man can possess. All we can do is to respect both points of view.'[8]

The Times, in the leading article already quoted, described the Commission's work as 'the careful balancing of one factor against another in a context of exceptional intricacy', and nominated it 'the highest monument yet erected to modern Whitehall rationality'. That was not, presumably, meant to be read ironically. Something more was expected and indeed demanded of it than careful balancing of one factor against another. In the tradition of the best Royal Commissions and national commissions of inquiry (many of them less expensive and less ambitious in scope than Roskill), it might reasonably have been expected to push back the frontiers of existing thinking on planning and the environment, even if only by a few yards. The Commission did not hesitate to break new ground on other matters, in the large assumptions it made about public values and policies for its cost/benefit analysis. It did not then shy away from 'gifts of wisdom and prophecy', though some would have wished it had.

In general, though, the Commission's handling of these planning and environmental questions, though needlessly inconclusive, was more sensitive than its critics gave it credit for. Its handling of the cost/benefit questions in the final stages, however, was not nearly so sophisticated. At the Stage V hearings, the Research Team's economic assessment of the four sites had been severely put to the test. The county councils, the resistance associations and most other bodies that had given evidence had criticised it and suggested modifications. Although some of the more far-ranging modifications had proved impossible to fit into the computer model, the Commission had accepted a number of them as valid suggestions. The Commission itself required some changes in the assessment and the Research Team too had modifications to make. Consequently, the Research Team was kept in being after the conclusion of the public hearings to revise the cost/benefit assessment. This was not at all what most of the bodies that had given evidence had expected. One witness at Stage V, Professor John Parry Lewis of Manchester University, said in a letter to *The Times* that it seemed rather like the expert witnesses in a court of law being invited into the jury room to play a part in the verdict. The Research Team was being allowed a second bite at the cherry that was denied to the experts who had criticised its first assessment. But possibly this was less of a lapse of justice than an ambiguity over the role of the Research team within the Commission's scheme. At the earlier stages of the inquiry the Team had been, as it were, a prosecution witness presenting a technical case which the Commission itself neither agreed nor disagreed with. After the conclusion of the Stage V hearings, however, the Commission regarded the Research Team more as its own technically-skilled servant: no longer a prosecution witness, it was now a sort of friend of the court. It was a situation not entirely without precedent, but it caused some raised eyebrows among the organisations that had appeared at Stage V.

There was plenty to surprise them in the revisions to the cost/benefit analysis. The modifications incorporated an element of doubt into the value of savings in travelling time, a revision of the passenger traffic forecasts, new assumptions about air traffic using Manchester and Luton Airports and new construction schedules for the airport's first runway. These revisions did not change the overall placings of the sites. One of the two sites to the north-west of London was bound to be the favourite, once the two large assumptions had been made that the airport would cater for passengers from all over the country and not only from the London region, and that the time-savings of these passengers, however small, should be given a monetary value and aggregated as one 'cost' of the airport. Some of the 'costs' of the third

London airport were clearly more debatable than others. The least debatable were the construction costs. Here, the revised cost/benefit assessment surprisingly gave an additional £7 million penalty to Foulness and a £17 million advantage to Cublington, which now became significantly less expensive to build than the offshore site. The Commission explained that this important change resulted from a rescheduling of the construction work at the two sites so that, for example, the expensive earth-moving operations at Cublington could be left until after the first runway was in operation and the airport had begun earning revenue. The revised assessment also awarded Cublington considerable cost advantages in both 'passenger–user costs' and in 'road capital' (where from being the most costly of the four sites it now became the least costly). The net result of these changes was that Cublington now became economically the best site on every major cost factor. It would be the cheapest in its demands upon the nation's capital resources (construction costs for the airport itself and for roads, rail links, etc.); it would be the least expensive in operational costs (airport services, airspace movements, user costs and so on); it would be the least costly in the penalties that it imposed on society as a whole (the effects of noise and on agriculture, public buildings and so on). It was all suspiciously conclusive.

A number of other unsatisfactory features of the revised cost/benefit analysis were examined in an article by Professor Peter Hall of Reading University published in *New Society*. It was in this article that Professor Hall described Roskill's task as 'a heroic attempt to extend the field of rational, balanced socio-economic inquiry into a very difficult area of decision'.[9] However, he went on to question not only the effect of the revisions and the comparative weighting given to different elements of the assessment, but also the way in which some of the basic data had been assembled:

'The commission has proved that Foulness bears a heavy cost penalty in those things that can be quantified, though the sums rest on shaky foundations. However, the commission has not proved beyond reasonable doubt that Foulness would be costlier in direct commitments to the national exchequer; and, indeed, it is almost certain it would not. The commission has shown that the case for Foulness on planning grounds is not as clear-cut as it seemed, but too much of its counter-case rests on the assumption that the behaviour of air travellers cannot be controlled and that, in particular, the growth of Luton would be inexorable. The commission has not taken account at all of the long years of future history outside the span of its cost-benefit analysis, in which the Roskill airport would surely still exist as a monument to the rightness or the folly of their decision.'

The considerations that the cost/benefit analysis had left out of account were remarked upon by many critics, including WARA which, within three weeks of publication of the Roskill Report, had produced a detailed commentary from its technical and legal experts. WARA detailed sixteen items left out of the cost/benefit analysis, including the benefits to other regions of increased air traffic through airports outside the south-east region, the effects on employment, commuter travelling, the loss to London's overspill of the Milton Keynes city, as well as damage to the amenity value of the countryside, to ancient buildings and established communities affected by noise. WARA concluded that 'the Commission has failed to recognise the real limitations of cost/benefit analysis and the analysis as such fails to put the economic issues as such into their proper perspective'.[10]

Among those who rejected the results of the cost/benefit analysis was Professor Buchanan who, in his minority report, confessed that 'the analysis has caused me more worry and heart-searching than anything I have previously encountered in my life'.[11] He went on to describe how he had stood in awe of the intellectual qualities and the dispassionate approach of the Research Team which made him aware of his own deficiencies, but had also made him anxious. 'I became more and more anxious lest I be trapped in a process which I did not fully understand and ultimately led without choice to a conclusion which I would know in my heart of hearts I did not agree with.' This confession must have struck a chord of sympathy with many people, if not the majority, who had been in any way involved in wrestling with the cost/benefit analysis in all its intricacies. But Professor Buchanan rejected not only the recommendation of his fellow-commissioners, which he felt was excessively based on the results of the cost/benefit analysis, but all except one of the twelve chapters of summary of evidence. His colleagues said that they learned of this with 'dismay and concern'.

There had been a parting of the ways between them and Professor Buchanan at a much earlier stage than even they had appreciated. *The Times* described the difference between them as 'that between the fox and the hedgehog in the Greek proverb. The fox knows many things but the hedgehog knows one big thing. The big thing that Professor Buchanan knows is that "the land of the country, after its people, is its most precious asset". '[12] Perhaps the Commission was too fox-like in some of its subtleties, but could one really describe Buchanan's professional horizons as being limited like a hedgehog's? After all, here was Britain's best-known planner, a man of vast experience in public service and in transport and urban planning. His note of dissent was couched in the most splendid prose, admired even by the advocates of Cublington airport. It was partly an expression of personal faith, but it was the faith of a learned man. Like Martin Luther, he had been sorely tried by the detailed scholastic arguments on all sides and was forced to test his intuitive faith in something more positive by going back to his Bible. Professor Buchanan's Bible was the 1947 Planning Act, which embodied the ideals of men of similar faith like Patrick Abercrombie and Raymond Unwin, whose concepts of the green belts and the 'open background' around London were used by Buchanan with such effect in his minority report. There was no agnosticism about his view of the planning and environmental issues raised by the comparison of the four airport sites:

'I think this case is as critical a test of our sincerity in respect of two related matters as we are likely to be confronted with during this century. The first is our determination to plan comprehensively and with social purpose the commercial and other developments that modern life requires, and the second is our concern for the environment of our not-very-large and densely populated island . . . I do not believe there is evidence available from any country in the world to show that sensible regard for the environment does in fact lead to economic disbenefit, indeed I would say that the reverse is true and that there is overwhelming evidence to show that concern for surroundings, covering everything from better houses to better schools, offices, and factories, and improved access to open spaces, invariably brings rewards in the long run.

'The choice of Cublington would be a grievous blow to conservation policies. It is not merely that there would be a direct setback in the area influenced by the site, even more serious would be the general sense of disillusionment that would come to every person and organisation labouring in the conservation movement, and come just at a point in time when the urgency of the subject becomes daily more apparent. On the other hand a decision which conceded the importance of the environment (as would be the case if Foulness were chosen, even allowing for the losses involved) would be an event of great significance for the future of Britain. It would show that this country at any rate, in spite of economic difficulties, is prepared to take a stand. I believe such a declaration of attitude would redound to our credit in more senses than one.'[13]

Professor Buchanan's personal declaration of faith was well timed. At the time of publication of the Roskill Report, the pace of environmental concern seemed to be quickening almost daily. This was partly a result of the setting-up of the Conservative super-Ministry called the Department of the Environment, which in the first bloom of its existence was attracting the attention and the hopes of everyone even vaguely concerned with environmental problems. There was an environmental movement abroad in the country, and Professor Buchanan's minority report provided this movement with a stirring, well-written manifesto. The majority Roskill Report, of course, provided it with a worthy test case.

So far this chapter has considered the Roskill Report on the merits of its own arguments and its recommendation. But it is possible to see it in another way. If one sets aside the arguments themselves, and takes instead the sheer weight of words devoted to each of the four sites, then the Roskill Report has a clear bias, less towards Cublington which it recommended, than towards Foulness which it rejected. Even many of the arguments themselves can be interpreted as being more concerned with the economic and planning penalties the nation would incur if it chose Foulness, than with the positive benefits of choosing Cublington or any other inland site. There are a number of reasons for this hidden bias towards Foulness. One of them is that the Commission had received more evidence in support of the airport at Foulness than in support of any inland site. This had been the case not only at the Stage V hearings, but also with the mass of written evidence and even at the local hearing at Foulness itself. However strongly the Commission had criticised or rejected most arguments in favour of Foulness, it could not help but admit that some of them were good arguments and ought to be mentioned in the final Report. Whatever else may be criticised in the Report, there is no question that Mr Justice Roskill's summing up was immaculate. Even arguments that the Commission rejected because it considered them beyond its competence to evaluate—like the plans for a joint airport–deep-sea port at Foulness or the use of an airport there as a means of revitalising the declining economy of South Essex—were faithfully recorded. Indeed, if one reads between the lines of the Report, one could conclude that the Commission half expected that its recommendation of an inland site was likely to be set aside in favour of Foulness and that it should consequently spell out the economic and social price of choosing Foulness. It is an interpretation of the Commission's possible motives that adds special point to the exposition of the basic dilemma about airport planning, quoted earlier in this chapter.

But did the Commission expect its recommendation to be set aside unless it chose Foulness? For some members of the Commission, though perhaps not for its chairman, the answer to this question is possibly 'yes'. Foulness was all along the popular and the political favourite. Indeed, some people in 1968 had seen the appointment of this elaborate inquiry merely as a way of giving official blessing to the novel idea of an offshore airport and, as a feasibility study of an offshore airport at Foulness, the Roskill Commission more than fulfilled the expectations of these people. Moreover, it was always understood that the ultimate choice lay with the Government. But which government? In the summer of 1970, this became the all-important question for everyone concerned, members of the Commission as well as environmentalists. After the Commission had begun its public hearings at the Piccadilly Hotel, London, one of its members heard 'on the best possible authority' that Mr Heath, at that stage still Leader of the Opposition, had arrived at the personal conviction that the airport should go to Foulness whatever the Commission recommended. If there was any truth in the report, then the site of the third London airport was as good as settled once it was known on June 19 that the Conservatives had won the general election. It is a story that is not widely known and is, of course, difficult to verify: a Prime Minister does not advertise or confirm the fact if he has prejudged such a major planning issue on which a painstaking commission of inquiry is about to report. Yet the political circumstances in which the Roskill Commission reported back were, to say the least, unusually confused—as will be discussed in another chapter. All that can be said about the story with any degree of certainty is that it was known to at least two members of the Commission long before they began work on drafting the final recommendation and Report. If they also believed it, then it puts an interesting gloss on the idea that the Commission was as concerned to make a negative recommendation about Foulness as it was to make a positive one concerning Cublington. On the other hand, it is scarcely credible that Mr Justice Roskill himself could have seen the final labours of his three-year inquiry in this way. A judge may deliver a judgment in the knowledge that it can be reversed by a higher court, but not in the expectation that it will be completely upset.

X
THE PEOPLE'S CAMPAIGN

To the outside observer, Cublington's reaction to the Roskill recommendation must have seemed like the counter-attack of a carefully mustered and well-disciplined army ready, the moment the enemy showed his hand, to go out and do their stuff before the eyes of the world. The press coverage of events that weekend, December 18–21, certainly appeared to bear out an impression of well-laid plans calculated for their effect on the media.

In the Saturday morning's papers, reports of Mr Davies's announcement in the Commons vied with 'human-interest' stories from the villages. The *Daily Mirror*, in a full-page report, showed a large picture of two Stewkley children praying at their school's annual carol service: 'In a beautiful old village church, 100 schoolchildren clasped their hands and prayed "God save our village".' The Luton *Evening Post* carried a similar report and picture, but added: 'It will take more than prayers to save their heritage. And so tomorrow the villagers around Cublington . . . will declare war in a traditional way. On hills throughout the county, bonfire beacons will blaze the message to the world "we shall never surrender".' The carols and the planned bonfires were featured in all the morning papers. In the afternoon papers, however, it was another tradition of resistance that was being reported. 'ARMED SIEGE BATTLE CRY' said an *Evening News* headline, 'TO ARMS! CALL IN AIRPORT ROW' (*Evening Standard*), 'WAR OVER WING' (*Evening Post*). All went on to report how leaflets calling for active resistance and advising on the use of 'communal barricades', petrol bombs, shotguns, catapults, even bows and arrows, had been distributed to homes in the Cublington area during the Friday night by a group calling themselves the 'New Revolutionary Party'. All three papers carried the denunciation of the leaflets by Desmond Fennell, chairman of WARA, who described them as the work of 'madhats' and a 'hooligan element', but it was also reported that the police, even the Special Branch, were investigating. But, as Patrick O'Donovan commented in the following morning's *Observer*: 'It is uncertain how serious the threat is. The official Wing Airport Resistance Association, which really represents the whole area now, has disowned the group publicly and is privately uncertain whether this militancy helps or hinders their purpose. But this is not a part of England where society is rejected or the ordinary usages of democracy despised and the fact that it has happened is both astonishing and fair evidence of the public rage.'

Much of the popular demonstrations against the Roskill recommendation was, however, good-humoured. The Sunday papers also reported an inter-village 'Knock-out Competition' held at Wing on the Saturday afternoon, where one of the best attended side-shows was a coconut-shy with the names of the Roskill commissioners—excluding Buchanan—painted on the coconuts. If there ever was a possibility of the militant 'madhats' spoiling WARA's good press, it went up in the smoke of the bonfires of that Sunday night. Thirty-five prominent sites in Buckinghamshire and Bedfordshire had been chosen. In fact, some sixty fires blazed that night, ringing an area of some 300 square miles. Some were large conflagrations like the 30-footer at Stewkley, others more manageable affairs like the four portable fires that were carried to the top of Dunstable Downs, Beacon Hill, the hill overlooking Wendover and the National Trust's Ivinghoe Beacon. There they were lit, then carefully extinguished and brought down again. Desmond Fennell described the fires as 'torches for the environment', a good phrase for capturing the theme of the moment (these were the closing days of European Conservation Year). The Monday morning papers, on the other hand, saw them as 'flames of anger' and 'blazing protests'. In truth, the historical associations of a chain of beacons such as had warned England of the approach of the Armada, and the tolling of muffled bells in six villages (two of them against the wishes of the incumbents) had become mixed up with other historical associations. At Stewkley, where the congregation trooped out of the church to sing around the fire while a band played, they also burned Mr Justice Roskill in effigy. There were similar echoes of November the Fifth at other villages. Bonfires were to have an important symbolism in later events.

Such was the impact of the local reaction to Roskill that anyone tuning in to BBC's Radio Four on that Monday morning would have heard reports or at least mention of it on the South-East news (6.50), the main news (7.00), 'Today' (7.10), 'Thought for the Day'

(7.45) and again on the South-East and eight o'clock news. Only the weather forecast did not mention Cublington. It certainly seemed like the response of a finely-tuned publicity machine. But the truth was otherwise. In the autumn, before the press had leaked the fact that the Roskill Commission was to report before Christmas and likely to recommend Cublington, the airport question had effectively gone off the boil. The last real contact that most of the people had had with the Roskill proceedings had been the local hearing in Aylesbury of well over a year before. The Stage V hearings at the Piccadilly Hotel had been over most people's heads. There had of course been fund-raising events, but by the time that these had dragged on through a second summer there was little enthusiasm for still more in the late autumn. Some WARA village organisers had had difficulty in getting together teams for the 'Knock-out' competition on December 19 and in interesting people in other events. There was also a considerable body of local opinion which had shown little emotion over the airport before the Roskill recommendation. These were the people who had felt that if 'They' wanted the airport at Cublington, ordinary people were really wasting their time in trying to stop it. Getting these people to turn out in large numbers for the 'events' of early 1971 was a major achievement.

For its part, the WARA leadership realised that with the Roskill choice of Cublington the £50,000 it had invested in legal and technical representation had been spent, if not in vain, then patently without obtaining the desired result. Perhaps, it was argued, too much had been spent on lawyers to the neglect of other professional services, like public relations and national advertising. Earlier in the year, in fact, the WARA committee had debated a proposal to employ a public relations firm. The balance of argument had been swayed by the offer of £1,000 as part payment for a PR firm from the Wings off Wing group (the WARA group in Weedon, Whitchurch and Oving). And so, during the summer an American firm of public relations consultants, Burson-Marsteller Ltd., was hired for three months at a special rate of £3,300—the deal to which some journalists later looked for an easy explanation of the success of the whole campaign. By the time of the Roskill recommendation, however, the three months had expired. Although Burson-Marsteller continued to do some useful work on a nominal retainer basis (it arranged two successful press conferences), its contribution to the campaign as a whole was really of marginal importance. The chairman of Burson-Marsteller, Claud

Simmonds, described its contribution: 'We sent a questionnaire to all candidates in the last election, then organised mailings to all MPs and Parliamentary reporters once the report was published. In particular, we advised on MPs and peers who might be sympathetic to the cause. But the information we circulated was drawn up by WARA, and the real political organising was done by an all-party group got together by local MPs... The association could have handled the whole campaign itself, and it would have won, but it wouldn't have done it half so well.'[1] WARA also received professional help from three advertising executives who volunteered their services, designed three full-page advertisements which appeared in newspapers in February 1971, and even tried to raise money to buy the space. Again, the effect was marginal, as one of the trio admitted: 'the movement had gained a considerable amount of momentum by the time [the advertisements] appeared, and they weren't conspicuously good ads anyway'.[2] Events were to show that WARA had no real need of an intermediary in its relations with either press or public, apart from its own part-time press officer, John Flewin. His freelance agency in Bletchley became the clearing house for plans from Cublington and inquiries from Fleet Street—though the national newspapers later established direct contact with some village groups.

In the weeks following the Roskill recommendation, the effective organisation of WARA underwent a radical change. The executive committee, which nominally at least had directed WARA policy in the early stages, ceased to provide the impetus for the local campaign. Its leaders, and especially the chairman, Desmond Fennell, gave much of their attention to the organisation of the all-party parliamentary lobby against an inland airport. The local campaign became the responsibility of Bill Manning, previously organiser of the 'grass-roots' petition and by now vice-chairman of WARA. His main task was to keep some sort of central control over the 'events' which were being suggested and planned by village activists—to give the WARA *imprimatur* where necessary, to arrange for construction work, painting, funds and so on, to space the events out to avoid congestion and to lay on a bush-telegraph network of communications. By the day of the Roskill recommendation, this last service was sufficiently organised for one call from Bill Manning made at lunchtime on December 18 to lead to something like 200 contacts being informed by the evening of the dates fixed in January for the WARA rally and the 'roll-on' of farm vehicles. The contacts ranged from parish meetings to MPs and reached geographically from St Neots to Stadhampton and from Banbury to the heart of London. Contingency planning of 'events' in case Roskill did choose Cublington had begun as early as August, when some three or four events were mooted—the rally, the 'roll-on', the bonfires and 'Operation Green Man', a scheme planned by the militants of the Home Defence Group, but never carried out, for a string of roadblocks around the Cublington area. The Home Defence Group, incidentally, denied all knowledge of the petrol-bomb leaflets distributed on the night of December 18, though this operation was in fact carried out individually by three of the group's members. Of the other events on that first post-Roskill weekend, the Stewkley school carol service was an annual fixture with ostensibly nothing to do with the airport, although the press with cameras and notebooks crowded into the gallery of the church found a special significance in this item of normal village life. The 'Knock-out' competition was planned as a fund-raiser and morale-booster before the date of the Roskill announcement was known.

The Cublington campaigners had built up a head of steam in the weekend following the Roskill recommendation and it continued to rise over the Christmas period. Although they claimed that theirs would be a 'traditional' Christmas unspoilt by Roskill, on December 23 they announced plans for even bigger demonstrations in the New Year—the 'roll-on' of farm vehicles around the airport site and the rally at which they expected to attract some 5,000 people. As important, however, was the fact that by Christmas 160 MPs had signed the 'no inland airport' motion that had been put on the Order Paper on the day of the Roskill announcement.

Terence Bendixson offered a nice observation on it all in the Christmas edition of the *New Statesman*: 'The oddity of Cublington is that 160 MPs have already sprung to its defence, thus lending a solid constitutional strength to the television entertainments being carefully engineered by the ex-urbanites and farmers of North Bucks, which nicely underlines the point that America is not automatically the model for Britain. A cultural gap as wide as the Atlantic divides the two peoples and nowhere is this more apparent than in attitudes to the land. But if the whole Stanstead/Cublington furore is very un-American, the assumption underlying it—that cities are hellish and countryside lovely—is in the mainstream of English thinking . . . The importance of Cublington lies in its impact on the imaginations of the English—the threat it represents to the goal shared by many, particularly in the opinion-forming classes, of a house in a village in "unspoilt country". It is this phenomenon that makes any inland site for a third London airport so politically tricky. What government will knowingly involve itself in driving barricaded farmers and shotgun-waving squires from their homes in full view of the zoom-lens audience, particularly when there is the makeshift option of Foulness?'[3] Although Mr Bendixson's view of the Cublington taskforce may have been fashionably inadequate (ex-urbanites, farmers and squires), he well summed up a major element in their strategy. Their appeal was to the imaginations of the English and these they captured without a doubt in the events of the New Year.

The concourse of vehicles rolled out on January 3 in freezing fog. It was the one notable occasion when they were let down by what came to be known as 'WARA weather', but the appalling conditions certainly heightened the visual effect. Some 300 vehicles turned out to drive the 28 miles around the perimeter of Roskill's proposed airport, an impressive array for the depths of winter—combine harvesters, muck-spreaders, tractors, lorries, land rovers, horse-boxes and fork-lift trucks. Fittingly the procession assembled on the tarmac of the old wartime airfield, where some of the vehicles had been parked overnight (under guard, for there were fears of sabotage). It was sent off by Bill Manning, who claimed it was the 'biggest roll-on of agricultural vehicles ever put on the road'. It took four hours for the 'roll-on' to trundle its way through the fourteen villages, where it was cheered by hardy crowds at the roadside. The loudest cheers came for some of the slogans born by the vehicles—'Spread Muck, Not Concrete', 'You Can't Milk Jumbos', 'Don't Foul Bucks, Foulness', and others whose message was personally directed at Mr Justice Roskill. On such a day they would have cheered almost anything. One of the most consistent cheer-raisers was the bus which collected villagers for the Sunday visiting at the local hospitals. Fears lest the bus be held up by the roll-on had been expressed by some of the older villagers. In fact, to the credit of the two organisers, Peter Warren of Wing and Don Bellingham of Aston Abbotts, the bus went through during the period when the concourse pulled up for soup and sandwiches, provided by the women of Drayton Parslow, Swanbourne and Mursley.

'It was a fine and impressive show of community feeling,' wrote Oliver Pritchett in the next morning's *Guardian*, 'children with their banners and roadside cheers and waves—rather like one of those nice old Ealing films where the people rise up against the bureaucrats and it all ends with a traction engine rally.' The same reporter discovered a fly in the ointment, a pro-airport driver who had come 'because his boss had told him to'. If he did, then he must have been the only reluctant participant. The weather conditions, in fact, made it an act of personal commitment for everyone else who took part and this is how it is remembered in the villages to this day—rather than as a jolly Ealing Studios-style carnival. People still tend to say: 'Old so-and-so—do you remember him on the roll-on?' However, the visual effect of vehicle after vehicle looming out of the fog was largely lost on the nation at large. Television news on both channels that evening was devoted almost wholly to the Ibrox Park football disaster in Glasgow.

The campaign was now gathering momentum. The roll-on was intended as the curtain-raiser to the following Sunday's rally. In the meantime, there was a solid demonstration of institutional opposition to Roskill. Bucks County Council called a meeting on January 7 of 100 local authorities and other bodies opposed to the airport. Unanimously they adopted a WARA motion affirming 'implacable opposition' to the choice of Cublington—'which would destroy for a doubtful short-term advantage the lives and homes of thousands of people together with a priceless piece of English countryside'. The emphasis of the meeting was as much on disruption of the people as on the environmental loss. Mrs Alice Beasley, chairman of Stewkley Parish Council, hit the dominant note when she told the meeting that 'the airport will be built on death, for it is plain fact, not sentimentality, to say that our elderly citizens are so worried now that the coming of the airport will kill them'. It was, however, the voice of one particularly elderly citizen that caught the attention of the press. The octogenarian Rector of Dunton, Father Sillitoe, asked how 'the destruction of great numbers of homes and the misery and the heartbreak involved' could be morally justified. He called the proposal to bulldoze churches and graveyards, in plain words, 'damned sacrilege', and promised, 'we will fight on the doorsteps of our homes, in the fields of our farms, at churchyard gates and at church doors'. This was a meeting of respectable county councillors and of representatives of bodies like the National Trust and the Council for the Protection of Rural England. Yet the determination to 'fight'—in some way or other—was its clearest message. Even the chairman of the Bucks County Planning Committee, Alderman Ralph Verney, reminded his listeners of the county's tradition of Hampden, Milton and Wilkes—'all fighters against tyranny in its various forms, all pledged to the inalienable right of the little man to live in his home unharassed by the tyranny of the public interest in whatever guise'. The determination to fight was probably there already: it was the thoughts of this 'tyranny' and of Roskill's 'materialistic philosophy' which fanned the flames of public anger. There was certainly anger at meetings like the one in Aylesbury and at the following weekend's rally but, unlike the narrow passions of personal anger, this collective emotion allowed room also for plenty of exhilaration and comradeship, as the rally proved.

The rally took place on a spring-like day as different as could be imagined from the 'roll-on' weather of the previous Sunday. It was held in the hangar-like building of the Equestrian Centre at Wing, which was thought to hold about 5,000 people at a pinch. Estimates vary considerably of how many did come. WARA said 10,000, the guesses of the press ranged from 8,000 to 20,000. What is a fact is that the Equestrian Centre was full to capacity, that there were a few thousand people outside it and that there were still people abandoning their cars on roads a couple of miles away and trudging across the winter fields when the proceedings were coming to an end. The hangar itself was lit by the arc lamps of several television crews, from Britain and abroad. There were loudspeakers to relay the speeches to the people outside and on a makeshift platform of two farm carts was assembled a battery of speakers which reflected how much of a national cause Cublington had become. Two Labour MPs were there (Reg Prentice and Miss Joan Lestor) and four Conservatives (Sir David Renton, William Benyon, Timothy Raison and Stephen Hastings, chairman of the anti-inland airport group in the Commons). Telegrams of support were read from other MPs. The main part of the proceedings was the speeches of these legislators which ranged in tone from cool dissection of the Roskill logic to impassioned calls virtually to revolution. Most of the MPs made it clear that they would vote against the Government if it tried to bring in a development order for the Cublington airport. There were speeches too from Desmond Fennell, from Alderman Verney and from Ernie Keen, the Stewkley farmer mentioned in Chapter V. Ernie Keen answered charges of class-bias in WARA by saying, typically, 'We aren't middle-class people, we are not upper-class or lower-class people—we're first class people.' The 'implacable opposition' motion, that had been brought from the Aylesbury meeting of a few days before, was carried by acclaim. It was later taken to Number Ten Downing Street by two local MPs, Messrs Benyon and Raison.

Reasoned opposition was only a part, even though an essential part, of the rally. There was a festive, almost a ritualistic, mood about it. One reporter compared it to an American political convention. There were banners everywhere and above the speakers hung a large cartoon by Zec, former cartoonist of the *Daily Mirror*, showing a jackbooted Roskill tramping down villages and scribbling in a book entitled 'The March of so-called Progress'. Whenever the name of Roskill was mentioned it was greeted with booing, while mention of Buchanan's name evoked loud cheers. Unknown to anyone, Professor Buchanan himself was in the audience. He had come out of curiosity at the WARA phenomenon and confessed later that he cheered his own name to preserve his anonymity. After the speeches, Cleo Laine sang the WARA battle song, a version of 'We shall not be moved' written by novelist Geoffrey Household:

> Brothers of England we are keeping it for you,
> Vale and tree and upland, keeping it for you,
> For they shall not slap concrete on our own dear land.
> We shall not be moved.

Then, Pied Piper like, Miss Laine, her husband Johnny Dankworth and a jazz band from Thame led the audience out of the hangar into the fields for the festive second half which began with the release of 3,000 gas-filled balloons. Each balloon carried a postcard, inviting the finder to send it as an anti-airport protest to the Secretary of State for the Environment. But hereby hung a problem. The balloons had been filled with helium gas the night before and, when the postcards were attached to them in the morning, they clung stubbornly to the ground. A much-publicised event was saved from going flat by the expedient of heating the balloons with a portable grain drier—and off they went into the blue. Attention then turned to a large replica of the Roskill Report (the imagined report, for the real one had yet to be published); it was made of large sheets of hardboard, packed with wood shavings, sump oil and other inflammable material, and it sat on top of a 6-foot pole. The burning of the Roskill Report was to be the culmination of the proceedings. The torch was set to it by Father Sillitoe who before he did so prayed, like a modern Elijah, 'that these inhuman and sacrilegious proposals be so absolutely rejected and reduced to nothingness as the flames of the fire shall reduce this copy of the Roskill Report'. The flames leapt upwards, a local brass band played the Funeral March and everybody cheered.

As a demonstration of mass feeling against the airport and as a political event, the rally succeeded beyond even the hopes of its organisers. It was the WARA happening that gave rise to the most comment from pundits in the press and the humorists too who found much to caricature in this assertion of rural feelings. In terms of the Cublington community, it was an achievement of a different order. It was, in the first place, an effort which called upon a whole army of talent. The main organisers were professional people—Jeremy and Isobel Smith-Cresswell of Drayton Parslow planned it, Norman Tennant of the same village designed the 'Roskill Report', it was made in the workshop of Bill Grace, the Stewkley builder, the hangar was decorated by David Stubbs and Colleen Burnett. The full list would embrace most village occupations. By this stage in the campaign detailed instructions and planning were superfluous: most people knew instinctively what to do. This was the case with the stewards who were told merely to be on hand in case of trouble. Not that trouble was specially expected—though there was a vague threat from a pro-airport farmer, or rather *the* pro-airport farmer, to dump a load of rubble in the car park. The main point of the stewards was to allay the anxieties of the many older, shyer and more nervous residents who were known to be attending what was probably the first and the last demonstration in their lives. It was one thing for them to stand at their cottage doors and cheer the roll-on vehicles as they went past, but quite another to join a crowd of thousands in a display of popular emotion that might well attract the sort of trouble-makers that older people associate with demonstrations.

Another wonder of the Cublington campaign by this stage was that the taste of it seldom cloyed with the national press and television, even after the extensive publicity given to such events as the roll-on and rally. Yet this was not the silly season, the midsummer doldrums when the press habitually has to make its own news. There was plenty of real news around, provided by Northern Ireland, by the Common Market negotiations, by the TUC, by a postal strike, by Ronald Biggs, Rudi Dutschke and Caroline Desramault. One can well imagine an exasperated news editor having told his Monday morning conference: 'Look here, we've run the antics of these airport people for a month now. It's about time we eased up or people will stop buying the paper.' If such a conversation did take place, it was ignored or overruled. Indeed the press and radio and television came looking for further stories after the rally.

The people themselves showed no signs of letting up either: after the rally they asked each other 'What next?' One answer was to intensify the placarding of the area with anti-airport signs. In the previous year, an audacious team of militants had painted 'No Airport' slogans on each of the three or four bridges over the main railway line between Leighton Buzzard and Bletchley. Whether this was the inspiration, or whether it came from the old-established signs beside railway lines like 'You are now entering the Strong country' in Hampshire and Dorset, the idea was born for one of WARA's best-remembered posters. Colleen Burnett, the young Stewkley sculptress who had painted most of the village's signs, was set to work. Her canvases were 24 × 6-foot expanses of hardboard on which she painted the now traditional Stewkley man with a pitchfork saying 'No'. These were erected in fields beside the track and must have been seen by thousands of travellers on their way to the Midlands. At about the same time smaller 'no' signs were mass produced and nailed to roadside telegraph poles throughout the area.

At the end of January, the activities of Cublington eased up a little. The full Roskill Report was published and people busied themselves in reading it and in sending their comments to the press, the Prime Minister and Parliament. WARA began immediate preparation of its own expert rejoinder and the Stewkley Action Committee prepared copies of Professor Buchanan's Note of Dissent for every Member of Parliament. The press was full of Cublington comment. There was also useful institutional support for the Cublington cause, reminiscent of that which had rallied to Stansted. On January 14, *The Times* published a letter of 'profound concern' at the Roskill recommendation from the presidents of the Civic Trust, the Royal Institute of British Architects, the Town Planning Institute and the Institute of Landscape Architects and from the chairman of the Council for the Protection of Rural England.

The inactivity at Cublington did not last for long. The Home Defence Group had drawn up its plans for 'Operation Green Man'. This was an elaborate scheme for stopping road traffic from entering the Cublington area and was meant to be a foretaste of the opposition that airport contractors' vehicles could expect. Portable barricades were to be thrown up on selected roads, all drivers were to be stopped and only allowed to proceed when they gave their support to the local resistance. If the police intervened, the operation was to switch immediately to another part of the airport site. Unlike some other Home Defence Group schemes, this one was submitted for WARA's approval and at first seemed to get it. But the atmosphere changed after the rally. Surely barricades would antagonise outsiders quite needlessly and perhaps cause accidents? 'Why stop the motorists from coming in?' asked Bill Manning. 'Why not invite them to come and see for themselves what we stand to lose?' The result was the 'Open Day', held on February 14 and organised very successfully by among others the Home Defence Group. Seventeen villages were thrown open to visitors, who were given free teas, shown typical village activities and taken on conducted tours. It is scarcely possible to say how many people came. The *Daily Telegraph* said a few thousand and reported that one home had received thirty callers and that there had been a half-mile traffic jam in Stewkley. It was in many ways the best of the events or happenings for it employed no gimmicks and it showed the people of Cublington simply being themselves in their own homes. It was not an entertainment for either the press or television, although both responded to it.

On the following day, the campaigners at last got the chance to take their protest to London. Eleven coaches, carrying 600 women and children (it was the schools' half-term), set out for Westminster. As the national postal strike was in full swing, they used the trip to deliver to the House of Commons the Stewkley copies of Professor Buchanan's minority report, one for each MP. But it was meant primarily to be a demonstration of ordinary people going to see their political leaders: they called themselves the 'Aylesbury ducklings', in contrast to the much-publicised Brent geese of Foulness. They also took with them charts showing the family network of Drayton Parslow, a family tree from Stewkley dating back to 1559, and a letter to the Prime Minister, penned in fine copper-plate. It was a good-humoured demonstration and, as far as the police were concerned, well planned in advance. The only hitch was that, with the postal strike on, the staff of the Palace of Westminster were reluctant to admit so much literature—or so many children—to the House. This difficulty was overcome by a tactful MP and an outspoken housewife. Yet the event did not gain much publicity and was the only Cublington event to fail in this way. On the following Sunday, however, there was comment on this fact from both *The People* and the *Sunday Express*, which asked: 'Must gentle village people carry on like anarchists or starlets to get publicity?' But still the ideas for more events continued to flow. On the coach back from Westminster, two Drayton Parslow housewives, Ginny Bates and Betty Bond, suggested a Shrove Tuesday pancake race along the lines of the annual event at Olney, farther north in the county. Despite the lack of publicity for the 'Aylesbury ducklings', the race went ahead. It was the third major event in nine days and it was fully organised from scratch within a week. There was added point to it, however, in that the House of Lords was to debate the Roskill Report on the same day. Seventy women took part in the race, organised into two teams, one representing Weedon, Whitchurch, Aston Abbotts and Cublington villages, and the other Stewkley and Drayton Parslow. The winner, Mrs Jean White of Drayton Parslow, crossed the finishing line to burst through a large paper replica of the Roskill Report and present her pancake to the Earl of Essex. The Earl, who was to make his maiden speech against the Cublington recommendation, ceremoniously received the winning pancake and put it in a specially prepared protest envelope. It was meant to symbolise the homely ingredients of village husbandry that would be destroyed 'if Roskill slapped down his vast pancakes of concrete'. It was the *Observer* that discovered a further irony: while the Earl of Essex lived in Bucks and was against an airport at Cublington, the Earl of Buckinghamshire was living in a council house at Southend-on-Sea, Essex, right next to Foulness, and naturally enough did not want a major airport there.

The Lords debate, as we shall see in the next chapter, produced not one voice in support of the Roskill recommendation of Cublington. All eyes were now on the Commons debate, a fortnight later. It was, as everyone in the villages knew, the crucial debate about the airport and it gave rise to a charming sign of Cublington's concern. Someone recalled that, while doing his National Service in the Royal Navy in Malta, he had seen all the houses around the Grand Harbour and Sliema Creek in Valetta lit with candles in glass jars along the pediments of their roofs. It was the Maltese celebration of Marian Year. Why should not the houses of the Cublington villages be lit with candles of vigil on the night of the Commons debate? The idea was put forward with some hesitation in view of the strong Nonconformist tradition of the area, but it endeared itself to Bill Manning who had been searching for a suitable event for that night. It caught on with surprising ease. The villagers who attended the debate arrived back in the early hours to find many houses still lit by the flickering candles. The Commons debate produced, as had the Lords, overwhelming opposition to the Cublington recommendation. All that remained now was for the Government itself to decide—Cublington or Foulness? That decision was to be delayed for another month and a half.

In this final stage of the campaign, there was evidence of a reaction against the success that the Cublington campaign had enjoyed. The small resistance group at Foulness—constituted then into the Action Committee against Foulness Airport (ACAFA)—held a press conference on March 3, at which it accused WARA of 'foul play' in using a high-powered Public Relations campaign and in putting up a propaganda smokescreen to hide the valid arguments in the Roskill Report. A number of papers published articles airing the view of James Curran, the pro-airport leader at Thurleigh, who accused the WARA movement of being a middle-class phenomenon. Mr Curran, who admits to being middle-class himself, briefly espoused the cause of an airport at Cublington but ironically enough was soon argued out of it by a strong deputation of working men from Stewkley and Drayton Parslow at a meeting at Ampthill. But perhaps the most damaging event of this period was a debate on BBC Television under the title 'Was Roskill right?'. Apart from allowing almost no say to the large audience of Cublington people who had been invited to the studio, it was the one notable occasion when WARA's supposed professionalism and steamroller tactics were very clearly not in evidence. Foulness won on points and Cublington retired with the hurt look of ordinary people bewildered by the technicalities of television production.[4]

On March 25, however, the Cublington people again showed their mettle. Two Ministers—Mr Michael Noble, Minister for Trade, and Mr Michael Heseltine, Under-Secretary in the Department of the Environment—came to Cublington as part of their tour to see all four sites for themselves. Mr Noble's chief, John Davies, came in secret a little later accompanied only by one MP. The Ministers arrived in a cavalcade of vehicles, surrounded by officials and county councillors and well protected by police. Their route was not announced in advance, but look-outs and bunches of banner-waving women and children were strategically placed to greet them. Also lying in wait was a 'bomb'. It lay in a hollow of a lane already cleared by the police outriders and above it was a rope across the road, bearing a banner: 'If you want a fight, pick this site'. It was a small plastic bag, containing a clock mechanism, batteries and what looked like a bundle of explosives. An innocent WARA look-out who chanced upon it just ahead of the ministerial cavalcade was promptly detained and questioned. The cavalcade stopped and the Ministers were warned. A police superintendent and William Benyon MP inspected the bomb. The latter, against police advice, picked it up and tossed it into a field. It did not go off.

Bomb disposal experts who arrived later discovered that it was a hoax: the 'explosives' were painted cardboard. This time the police came to Cublington in earnest and questioned a large number of suspects. But the Cublington bomber remains undetected to this day.

It was now a matter of waiting for the Government's decision, though there was little inclination just to sit back and wait. The tension was extreme and the spirits of local people fluctuated between optimism and intense gloom. Even when rumours of a reprieve began to flood the press, there was no inclination to hold precipitate celebrations. After two years of activity, they were determined to wait until their reprieve was official. As much to keep up their spirits as anything, another 'event' was staged at Stewkley on Easter Monday. It was partly a ladies' football tournament between eight village sides and partly a traditional fête at which teas, an Easter bonnet competition and sideshows drew in the crowds from the surrounding area. It was designed less as a publicity stunt than as a celebration of the comradeship that had been created by the anti-airport campaign. Significantly enough, it is this event which is being repeated on subsequent Easter Mondays to bring together again those who saw so much of each other while they were fighting the common airport enemy.

The news broke that Cublington had been reprieved a good two days ahead of the announcement by John Davies in the Commons on April 26. The people were prepared for the victory celebration and so too were the media. BBC Television decided to open its edition of 'Panorama' with film of the bonfire at Stewkley, which was to be an inevitable part of the victory celebrations—they even managed to get the lighting of the fire postponed to fit in with their plans for a live transmission. Yet for the people concerned it was a solemn, family affair. The celebrations began with a service of thanksgiving in Stewkley church, conducted by the Vicar, the Rev Paul Drake, Mr Warwick, the Methodist Minister, and the Rev Hubert Sillitoe, and then a torchlit procession led the way from the church to the bonfire, where hymns were sung. It is reliably reported that the pubs were empty until after the celebration, though there was free beer later. The Cublington victory was tinged with a weary relief that it was all over and with sadness that now another community would suffer. The first prayer said in Stewkley church was for the people of Foulness.

XI
POLITICAL VICTORY

The Foulness decision was announced by the Secretary for Trade and Industry, Mr Davies, in terms which resounded with the smack of firm government:

'As the Commission's report stressed, on environmental and planning grounds the Foulness site is the best, and the Government have concluded that these considerations are of paramount importance. In the Government's view, the irreversible damage that would be done to large tracts of countryside and to many settled communities by the creation of an airport at any of the three inland sites studied by the Commission is so great that it is worth paying the price involved in selecting Foulness.'[1]

It took the Government five months from the date that it received the Roskill Report to decide that safeguarding the environment from an inland airport and particularly from an airport at Cublington was 'worth paying the price'; its eventual decision was based on the recommendation of a special Cabinet committee, chaired by the Home Secretary, Reginald Maudling.

Once the decision was taken, Conservative Ministers naturally enough made capital out of it as proof of the Government's lofty motives. 'For the first time a government taking a major national decision has given pride of place to the environment,' said Mr Heath;[2] 'Foulness is the world's first environmental airport, of which we shall be proud,' said Eldon Griffiths.[3] Their claims to real concern about the environment may well be justified, but they should not be allowed to obscure the fact that the choice between Cublington and Foulness was every bit as keenly-fought a political issue as Stansted had been. And in the political campaign, WARA played a central and crucial role, despite protestations to the contrary by Ministers like Peter Walker. Certainly the association's critics thought so. In Parliament, in the press and among the members and staff of the Roskill Commission, there were many resentful voices raised against the way that WARA was alleged to have influenced the Government's decision. Their opinions must count for something.

It was about half-way through the Piccadilly Hotel proceedings that WARA's leaders realised that they were likely to lose their case before the Roskill Commission and ought to be thinking of how to win the subsequent political battle before Parliament and the public. At this stage, not everyone shared WARA's pessimism; even after the hearings had finished in the autumn of 1970, Bucks County Council was sure that it had won and that the Commission would recommend Foulness. WARA's first political move was the unworthy one of trying to ditch its allies to save Cublington. If the Commission had indeed turned its back on Foulness and would only consider an inland site, it was argued, then why should it not be the Bedfordshire site, Thurleigh, rather than Cublington? WARA's counsel, Mr MacDermot, argued along these lines in his final speech at the Piccadilly Hotel hearings—and he produced some respectable planning and social reasons why Thurleigh, which already had an operational airfield at the Cranfield Institute of Technology, was preferable to Cublington. But it was tactically a bad move. It immediately strained the hitherto cordial relations between WARA and the Thurleigh resistance group, BARA; it placed in jeopardy WARA's own support in the northern part of the Cublington area which would also be affected by noise from Thurleigh; and more generally it alienated people who ought to have been sympathetic to the Cublington case. This feeling was summed up by Mr Crosland's famous jibe in the Commons debate that 'WARA is in danger of giving the impression that it does not care tuppence for anybody else's noise and environment.' WARA was rebuked by its two new Conservative MPs, William Benyon (North Bucks) and Timothy Raison (Aylesbury), when soon after the general election it tried to enlist their help in pressing for Thurleigh. The MPs pointed out that, if Mr Justice Roskill did indeed recommend Cublington, its best hope of a reprieve lay in all-party pressure on the Government to reject *any* inland site for the third London airport.

The Commons committee against an inland airport was already in existence before the June 1970 election, but dormant. After the election it began mobilising support for instant pressure on the Government if and when the Commission recommended an inland site. Its new chairman was Stephen Hastings, the Conservative member for Mid-

Beds, its new vice-chairman Mrs Shirley Williams, a former Labour Minister and the Member for Hitchin, and its secretary was Mr Benyon. Thus all three of the Roskill inland sites were represented on the committee which also embraced both major parties and was led by two thoroughly experienced parliamentarians. However, much of the committee's drive—particularly after the Roskill Commission had reported in December—was supplied by the three new Tory MPs whose constituencies surrounded the Cublington site: William Benyon, Timothy Raison and David Madel. Their initiation in the political handling of the airport question went hand in hand with that of WARA itself.

The most urgent task before the association now was to broaden its political base. During the election campaign, Burson-Marsteller, the public relations firm it had retained for three months, had circularised all parliamentary candidates with the association's case against an inland airport. Before the election, WARA had been wary of taking any bolder political initiative which might again have provoked the sort of criticisms of party bias that had emerged in the confrontation with Robert Maxwell a year earlier. But, with the election out of the way, the executive invited the three new local MPs to become vice-presidents of WARA and, in the months ahead, enrolled a further ten MPs as vice-presidents (eight of them Conservative and two Labour), while retaining the two former Labour MPs, Mr Maxwell and Mr Roberts, along with Sir Spencer Summers who had retired as Tory Member for Aylesbury at the general election.

Another pressing need was to draw up a tactical timetable of future action, based on as sound an intelligence as possible of the Government's intentions with regard to the airport. The MPs with whom WARA maintained contact were a useful source of information, but for a quite considerable period the likely intentions of the new administration were inscrutable to all observers. Nobody inside WARA knew the story then current within the Roskill Commission that Mr Heath had already made up his mind in favour of Foulness. Yet there were signs that the Government was serious about its avowed concern to protect the environment from unnecessary technological damage: Mr Heath himself made an encouraging speech at the Guildhall in London in October. On the other hand, the early decision during Peter Walker's tenure as Minister of Housing and Local Government, to allow a major brewery development at Samlesbury in the heart of the Lancashire green belt, had brought protests from conservationists everywhere. When he spoke about the environment at that autumn's Conservative Party Conference, Mr Walker gave the impression to his WARA observers that he was more likely to show his environmental concern in slum clearance schemes than in rural conservation. The setting up of a new Department of the Environment under Mr Walker and its subsequent display of comprehensive concern for environmental problems were to come later. In any case, the decision on the location of the third London airport was not the responsibility of Mr Walker's new Department nor of any of its component Ministries. It was a Board of Trade responsibility which under the terms of the Conservatives' White Paper in October on the Reorganisation of Central Government passed to the other new super-Ministry, the Department of Trade and Industry, headed by John Davies. Mr Davies's political philosophy was known to everyone: industry must stand on its own two feet and 'lame ducks' could not expect to be helped along by public subsidies. So in the circumstances, would the Government regard an offshore airport at Foulness as the best possible environmental choice or as a potential economic 'lame duck'? Were the decision to be Mr Davies's alone, there seemed to WARA to be little question that the decision would be made mainly on grounds of economic viability. If anyone then had told the WARA leaders that Mr Davies would later overturn a Roskill recommendation of Cublington with a statement that environmental considerations were of 'paramount importance', they would scarcely have believed it.

There were other uncertainties. They concerned what became known in WARA circles as 'the machinery of decision-making', in other words the method by which the Government, having received the Roskill Report, would arrive at its own decision on the airport location. The problem was whether the Government would publish the report without comment, thus allowing time for Parliament and the public to digest and criticise the Roskill recommendation; or whether it would make its own decision in advance of publication, as the previous Conservative administration had done in 1963 when Julian Amery had written in a foreword to the Interdepartmental Committee's report that the choice of Stansted 'though not perfect seems to be the only suitable site', and as the Labour government had done in 1967 with the White Paper on Stansted? The question was a crucial one for WARA, for it meant the difference between fighting just the Roskill Commission's recommendation and fighting a decision already made by the Government. There were plenty of people around at the time who thought that they knew how the Government would behave but most of their predictions, especially those that appeared in the press, appeared to be based on guesswork and rumour. WARA decided that it must enlist the help of men with actual administrative experience of decisions similar to that before the Government, and it formed a 'think-tank' of such men from the Cublington area. It included two former ambassadors, Sir Ralph Murray and Henry Hohler, and a number of businessmen—Tom Grieve (former vice-chairman and managing director of Shell-Mex and BP Ltd), Sam Allen (a director of the *News of the World* and the *Sun*), David Kessler (publisher of the *Jewish Chronicle*) and Sir Gilbert Inglefield (architect, businessman and former Lord Mayor of London). But there seemed to be as many views as people in the 'think-tank' and WARA was left with the still uncertain prospect that it might well have to mount a campaign against a government *fait accompli*.

At about this stage—in the month or so before the Roskill Report was published—the parliamentary committee against an inland airport and the political lobby of WARA became to all intents and purposes a single pressure group. On December 2, the committee invited Desmond Fennell and Bill Manning, chairman and vice-chairman of WARA respectively, to address a meeting in the House of Commons. Mr Fennell went armed with a short and informative speech which he had had the foresight to have duplicated. In it he listed the vital statistics of the third London airport—the biggest international airport in the world, three times bigger than Heathrow, employing 65,000 people, handling 600,000 flights and 100,000,000 passengers a year, needing 15,000 acres and, together with its dependent urbanisation and other development, probably taking 75,000 acres in all. 'Those figures demonstrate,' he said, 'what in essence is the problem at the heart of Roskill's task. It is whether you consider this as the largest piece of transport investment undertaken in this country—i.e. the market approach as it is sometimes called—or whether, without ignoring the market considerations, this vast project can be made to further some major social object and pay proper regard to the environmental and social consequences of its development.' All copies of the speech were distributed and a request was later made for further supplies. A fortnight later, WARA's officers again met the parliamentary committee in the company of other anti-airport groups, including those from Luton and the Heathrow area, who showed some reluctance to commit themselves to any rearguard action in defence of Cublington—if the Roskill Commission recommended it, as by now most people believed it would. On December 18, Mr Davies announced the broad outline of the Roskill recommendation to the House of Commons, and on the same afternoon an 'early day' motion opposing any inland site for the third London airport was tabled with the signatures of 160 MPs from all parties. In the evening WARA held a press conference in London—its last function to be arranged by Burson-Marsteller—with half a dozen MPs on the platform. Later many of the same MPs appeared on the platform at the WARA rally in Wing.

The first phase of the battle had been won. The Government had not announced its immediate support for the Roskill recommendation. The press had given overwhelming attention to the WARA case and to Professor Buchanan's note of dissent, which amplified the WARA point that an airport at Foulness could be socially advantageous to the people of south Essex, while a Cublington airport would bring only social and environmental disaster. The association at this time was already embarked upon the series of popular demonstrations, related in an earlier chapter, which were to keep its case before the public eye right up to the time of the Government decision. But when would that decision be made and which way would it go? Desmond Fennell put this point to a senior Conservative Minister whom he met socially, and received the reply that Cublington would probably be all right in the end, but that they had to keep up the pressure. On January 12, the Leader of the House of Commons, William Whitelaw, was the guest in the BBC radio programme, 'This is Your Line'; Mr Fennell phoned in to ask whether the Government would allow adequate time for debate before it made its decision. Mr Whitelaw's answer encouraged William Benyon to put down a question to Mr Whitelaw in the House. The answer was now more explicit: there would be a full debate in both Houses of Parliament before the Government took its decision.

WARA now set about the laborious task of supplying all members of both Houses with a full exposition of the association's reply to the Roskill Report. The full Report was published on January 22. Within five days it had been read by the association's technical experts and the members of the 'think-tank'. The following weekend their comments were embodied in a 15-page refutation of Roskill, drafted by Niall MacDermot, who had flown over for this purpose from Geneva, where he was based as the secretary-general of the International Commission of Jurists. The WARA reply to Roskill was published less than three weeks after the Commission's Report. The problem now was—at the height of the postal strike—to get a copy of it to every Member of Parliament. Here the WARA leaders encountered the same difficulty that was faced in the same week by the 'Aylesbury ducklings'—the ladies who brought every MP a copy of the Buchanan minority report. The staff of the House of Commons had been instructed to accept delivery of only one copy of each leaflet or pamphlet (it would presumably be available to Members in the Library of the House) and to relax the rule only if the literature related to parliamentary debate taking place in the same week. The Commons debate on Roskill was as yet unscheduled. In any case, if WARA could somehow overcome this restriction, its reply to Roskill stood more chance of being read by MPs than if it were delivered with all the other mail that would flood into the House once the postal strike was ended. It looked as if 630 WARA members would have to present themselves at the St Stephen's Entrance, each bearing an individually-addressed envelope inside which the pink-covered WARA booklet was hidden. The problem was eventually overcome by WARA's local MPs taking the booklets into the House and distributing them—along with the copies of Buchanan—from their offices. The manoeuvre was certainly worth the trouble that had gone into it and its success could be gauged in two ways. During both the Lords and the Commons debate on Roskill, members of the association in the public galleries were delighted to hear points from the WARA booklet being used in the debate and to see splashes of pink among the papers held by backbench MPs. There were also back-handed compliments to WARA in the complaints that 'not even the postal strike had been able to hold these people back'. Indeed, a Labour MP, Gerald Kaufman, had already asked the Speaker if Members could be protected from the barrage of WARA literature.

The first debate on the Roskill Report was arranged for the Lords rather than the Commons. It was a wise precaution on the part of the Government's business managers in view of the mangling that the Stansted proposals had received in the Upper House three years before. WARA had not neglected the Upper House. After his address to the Commons Committee in December, Desmond Fennell had been invited to speak to a similar meeting in the Lords and, when it came to the problem of trying to breach the postal strike restrictions, WARA found the Lords a rather simpler proposition than the Commons. A pile of the pink booklets was deposited without much difficulty on a table in the Lords' Library where interested peers would collect them. Few people, it seems, attempt to lobby the Lords as WARA did. The debate took place for two days at the end of February and was opened and closed by Lord Molson, chairman of the Council for the Protection of Rural England. It was a wide-ranging, non-party debate in which spokesmen for the Government and Opposition were careful not to commit themselves. The star performer was the Earl of Essex, whose maiden speech in defence of his native Buckinghamshire was praised by the Marquess of Salisbury. The remarkable thing about the two-day debate was, as Lord Molson observed, that the peers could not remember any other inquiry commission 'who have made a recommendation that has subsequently been debated in either House and not a single speech has been unequivocally in support of the recommendation made'.[4]

There was almost as little support for the Roskill recommendation in the one-day Commons debate at the beginning of March. Forty-one members took part and not one wholeheartedly endorsed Cublington as the best site for the third London airport. By the time that the debate took place, the number of signatures on the early-day motion against an inland airport was approaching its final figure of 218. This group embraced a wide and motley variety of party and constituency interests. There were MPs from the Roskill sites, veteran campaigners against aircraft noise, some who felt that a four-runway airport was probably unnecessary anyway, Members from constituencies in the Midlands, Lancashire and Yorkshire who wished to see local airport services expanded and saw little chance of this happening if the new airport were built at Cublington and, of course, conservationists. One Conservative is reported to have signed the motion solely out of concern for the National Trust collection at Waddesdon. Most of the 218 signatories were, in fact, Conservatives some of whom had indicated that they would defy the party whip if the Government sought to implement the Roskill recommendation. The fact that there were fewer Labour Members among the signatories was probably due to their party's preoccupation at the time with forthcoming legislation on both the Industrial Relations Bill and Common Market entry and, perhaps also, to the suspicion of class bias that hung around the conservationist lobby and in particular WARA.

On April 26, after the Government had taken further soundings of opinion at the four Roskill sites, it announced that the airport would be built at Foulness. WARA had won, but was it a victory in which the association's contribution had been decisive? It is impossible to say until the Cabinet Papers of the first half of 1971 are made public or until one of the Ministers involved in the decision publishes his memoirs. At a press conference on April 26, Mr Walker dismissed the WARA effort as playing no part in the Government's deliberations and indicated that his own Department of the Environment had been able to exert decisive influence over the final decision in a way that none of the old Ministries could have done. Yet a government as concerned as the Conservatives were with getting the best possible return on investment in the public sector, could not have afforded to view the third London airport decision solely or even predominantly as an environmental issue. It had far too great an importance for the future of the country's aviation and trade and for policies concerning the distribution of wealth and jobs in Britain's regions. This was how the Secretary for Trade and Industry, Mr Davies, saw it—at least on the evidence of the comparatively few public statements he made about the Roskill Report before April 26. The decision was, in fact, a very finely balanced thing. As that other Cabinet Minister had told Desmond Fennell, Cublington had needed to keep up the pressure. The pressure was considerable. With no voice in either House of Parliament supporting the Roskill recommendation and with 218 MPs and the majority of the press actively opposing it, what government would press ahead with such a decision, unless it were very sure of itself? But the Cabinet appears to have been divided and was certainly not anxious to precipitate a prolonged struggle with either its parliamentary supporters or the country at large when there were other, more important issues on which it might have to fight, like industrial relations or Europe. This is not to say, of course, that the Government's decision was devoid of principle or policy and that it was swept off course, as some of its critics said, by the concerted lobbying of WARA and the parliamentary committee against an inland airport. But, without that pressure, the final decision on the third London airport could well have been a different one.

XII

WARA'S 'DISREPUTABLE BROTHER'

What would have happened if the Government had chosen to go ahead with Cublington airport? Would the people there have given up their resistance and bowed to the decision? Or would the anti-airport campaign have continued and perhaps entered a new phase of forced evictions, of people sitting down before the bulldozers, or barricades and worse? Were there really people at Cublington who were contemplating guerrilla action in defence of their homes and environment?

These questions were asked with increasing seriousness in the last months of the Cublington affair. They were prompted partly by the vigour and determination of the WARA campaign, partly by all the fighting talk—about the John Hampden tradition, for example, and about resisting the airport if necessary by force—but most of all by a series of incidents which showed that some people wished the fighting talk to be taken literally. These people were a small minority in the anti-airport movement, whose schemes for direct action were in any case often only a small part of the effort that they themselves put into the peaceful demonstrations, and their activities were roundly condemned by the WARA leadership. But they cannot be ignored. As well as earning Cublington a bad name in some quarters, they were a symptom of the desperation with which many local people viewed official handling of the third London airport affair.

Most of the incidents and plans for direct resistance originated in Stewkley. That was to be expected in view of the village's character and its plight as the largest community to be threatened with destruction. This does not mean that all the militants were Stewkley people nor that the majority of the villagers either knew or, even less, approved of the aggressive plans being hatched in their midst. But Stewkley is where this aspect of the airport battle first appeared in May 1969, two months after publication of the Roskill short-list. At a meeting of the Stewkley Action Committee, it was suggested that a 'last-ditch' document should be circulated in the village, to be signed by 'all who were prepared to stay put whatever the Roskill Commission might recommend or the government decide—i.e. those prepared to refuse to leave their homes short of superior brute force being used to eject them'. A few weeks later there was a proposal for a 'Home Defence Group register' to serve roughly the same purpose as the 'last-ditch' document. The Action Committee treated both proposals with caution and agreed to them on condition that they were approved by WARA and a lawyer. That is the last mention of either in the Stewkley Action Committee minutes. However, a rough canvass of the village was made at about this time which showed that some 40 per cent of the householders had said they would not leave their homes unless forced to do so. Formal declaration of these intentions was made in the letters sent to the Roskill Commission, many of them by way of the County Council, and in oral evidence to the Roskill local hearing at Aylesbury.[1]

But the idea of a 'Home Defence Group' survived and in the autumn of 1969 a group of that name was formed. It was to be a nucleus of 'local people dedicated to fight the airport by direct action'. Its 'membership' consisted of five from Stewkley and five from outside the village, but they were to make contact with like-minded people throughout the area and to call on their services as required. As is often the way with such clandestine groups, one of the first tasks it put its mind to was a declaration of aims. It was a single-page, eight-point document, which included the following:

'The Home Defence Group is formed to promote popular action against the Cublington project. Broadly based militancy could (*a*) perhaps influence the final verdict of Roskill and without any doubt 'stepped up' popular action could influence Parliament and might be a major factor, (*b*) hearten resistance in refusing compensation and resisting eviction at a later stage.

'The Committee must be capable of implementing any threats it may make, or promises of help, otherwise these are counter-productive.

'HDG is not a publicity body and its purpose is not to contribute to the national argument. It is able to secure publicity of the right kind for its own demonstrations or incidents which it will mount. This does not prevent members writing articles or letters as they think fit.

'Nothing prevents members from working for WARA. HDG can be regarded as WARA's disreputable brother. Ties of blood must be stronger than any difference of opinion. Whilst HDG will keep its independence, it will listen to WARA views and it hopes for discreet support.'

The relationship between the group and WARA was less dramatic than that of a 'disreputable brother'. Most members of the WARA executive and village committees knew nothing of the militants other than what they later read in the newspapers. The few WARA leaders who did know tended to write off the HDG as 'Dad's Army', an amateurish, rather aged Home Guard caper. For its part, the Home Defence Group believed that it was fulfilling a need that WARA had missed. 'We were trying to stir up popular discontent,' one of the militants said, 'to make people aware of the danger. We tried to keep things on the boil. WARA hadn't got the contact with people. We realised early on that the strength of the campaign was in its disunity, not in its unity. There were so many people discontented with WARA that this provided the way to get things going. Once they were going, of course, we had all sorts of suggestions. People were always suggesting capers. We turned down a number of proposals for action because they were impractical or just silly. However, our main trouble was that nobody ever did what they were told. They went off on private ventures, even doing their own private planning.'

This happened with one of the first Home Defence Group ventures. It was the plan in April 1970 to placard the homes of Mr Justice Roskill and Mr Peter Masefield, chairman of the British Airports Authority, with 'no airport' signs. It had been agreed by the group that no lasting damage was to be done to property and nothing done that could identify the culprits with Cublington. The party that went to Sir Eustace Roskill's country home near Newbury—three men and a woman in two cars—kept to these arrangements and prided themselves on doing the deed and getting away in seven minutes. The other party had the more difficult task, for Mr Masefield's Reigate home is in a well-lit part of the town. In an excess of bravado the two men assigned to the job took paint pots and brushes and were clad in polythene sheeting over their lounge suits. They daubed the gates of Mr Masefield's house, nailed the placards over the paint and drove away, dumping the paint, brushes and polythene on the way. The slogan they had painted was: 'Not Cublington-Wing'. WARA was understandably furious. The Stage V hearings had just opened at the Piccadilly Hotel. Public attention had already been drawn to the Cublington militants by a printed letter distributed in Stewkley, Cublington and Dunton a fortnight before. It called for 'resistance to the finish' and bore the imprint of the Home Defence Group. Its advice was 'Remember that it is hard enough to evict a single family from its home. How much harder to evict the whole population of the vast area to be covered by the concrete of the Runways and Approach Roads! Be resolute, and the eyes of the whole country will be on you. Public opinion in the whole country will approve your stand. It will be impossible to use police and troops against you.' The group did in fact stage a mock eviction in Stewkley and had plans for an eviction show to tour the villages.

A later and more ambitious venture in 1969 achieved no publicity and came near to farce. Someone produced a journalist from South Buckinghamshire who said that he could get a story about the group into continental magazines like *Paris-Match*, whence it would filter back into the British press. For his benefit, the group laid on a 'military manoeuvre' in a large copse near Stewkley. The journalist was to be brought to Stewkley and then taken by a roundabout route, involving a change of cars, to the woods. There he would see exploding bombs (a mixture containing sodium chlorate weedkiller in plastic bags, but quite effective), half a dozen men and women practising with shotguns and old rifles, and an operational HQ. In the HQ was a chart that purported to be the Home Defence Group's chain of command—it would have done credit to a staff college, said its author. Two things went wrong. One of the cars failed to turn up and the journalist was left kicking his heels in a house in Stewkley. Then, his host received a telephone call which he said was a warning from an acquaintance in the Buckinghamshire Constabulary. The journalist left disconsolate and a messenger was despatched to the copse, where the 'troops' dispersed and an unsuccessful attempt was made to explode the bombs. One of them went off three months later when workmen were felling trees. The only other incident in which the Home Defence Group as a body was involved was the planning of Operation Green Man, the plan to put up selective roadblocks around the Cublington area. This was vetoed by WARA, as noted earlier, and instead the HDG helped organise the successful but quite uncharacteristic 'Open Day' of the villages.

As a group, they had failed to pull off their great publicity coup. This was to come as a result of the individual activity of two or three members. In the early hours of the day after the Roskill recommendation, copies of a leaflet were put through letter-boxes and nailed to trees and notice boards in the airport site villages. The leaflet took up the theme of the Home Defence Group letter of the previous April. Villagers were advised to conserve supplies of fuel, food and water and to open up old wells where possible. They were told to be ready to erect 'communal barricades'; to put by protective clothing like motorcycle helmets, visors and goggles, and dustbin lids; and to be prepared to use weapons including shotguns, air rifles, staves, catapults, bows and arrows and petrol bombs. The leaflets contained a diagram showing how to make a petrol bomb (taken, without permission, from a handbook for revolutionaries by Tariq Ali). Some of the leaflets were signed by 'the New Revolutionary Party'. As noted earlier, they received dramatic publicity. Another freelance incident was the brilliantly executed planting of the 'bomb' in front of the Ministerial motorcade during Mr Noble's visit to Cublington in March 1971. Finally, there was an advertisement put into the Personal Column of *Private Eye*, telling would-be volunteers to stand by for 'Japanese-style resistance' at Cublington.

Apart from rather double-edged publicity, the militants achieved very little. They were not a 'major factor' in influencing parliamentary opinion, indeed they may have caused some MPs to take a hostile view of Cublington opinion. They were laying plans for a last-ditch situation that never arose and that many people refused to believe could arise. The image of themselves that they fostered, of a ginger group or resistance 'command cadre' which would ensure that any force used by Authority would be met by comparable local force, seemed rather irrelevant and sensational. It conflicted with the image that WARA was fostering, of a community of peace-loving country people who deserved the sympathy and support of every law-abiding Englishman. However, the Home Defence Group and the individual militants did have something to show for their efforts. They attracted plentiful attention to the spirit of determination that existed in the directly threatened communities, not all of it censorious attention. They uncovered a great many people who were prepared to participate in direct action against the airport: some seventy people, for example, had agreed to man the roadblocks for Operation Green Man and many others could have been called upon in the event of the Government approving the building of the airport at Cublington. To this extent, the Home Defence Group did embody some of the impatience that local people felt towards the cautiously legalistic approach of the WARA committee. But the militants also caused a lot of anxiety among older inhabitants in the airport villages.

At the time when they were designing petrol bombs and planning how to set fire to contractors' vehicles should they appear, they were undoubtedly out of step with majority opinion in the Cublington area. As the Vicar of Stewkley and Drayton Parslow, the Rev Paul Drake, wrote in his parish newsletter at Christmas 1970:

'Regrettably, support for our cause will be lost overnight if any hotheaded "militant" is so irresponsible and foolish as to try to fight this battle for the minds of our fellow-countrymen with weapons of violence. To offer violence in the course of our justified protests would be a tragic mistake on at least three counts:

'1. Violent protest is un-Christian and therefore wrong in itself.

'2. Violent protest is counter-productive anyway and would achieve the opposite of its desired effect in the shortest possible time.

'3. Violent protest would be played up eagerly by the news-hungry media. Having splashed their front pages with banner headlines about "civil war in Bucks" they would have no difficulty in choking off all general sympathy for our plight by representing us as a mindless minority of hooligans indulging in the sort of destruction we had claimed to be protesting to the Commission about.'

Undoubtedly, the majority of Cublington people shared Mr Drake's view at the time. But what was the alternative? Would the people of Stewkley and the other villages, particularly Stewkley, have left their homes willingly to make way for an airport approved by Parliament? It is scarcely conceivable. They would have resisted to the end or very nearly to the end, but not with the tactics of the Home Defence Group. There would probably have been two strands to the resistance, one violent and condemned to failure and to 'choking off' public sympathy, the other passive and with much more chance of success. The violence would have been difficult to avoid. One or two farmers in the area had already made preparations unconnected with the antics of the Home Defence Group, although they made no show of these preparations while the Roskill recommendation was still being considered by the Government and Parliament. The author was shown a shed where the owner uncovered what he said were Army-surplus bazookas. They could well have been: this man was capable of anything. Elsewhere in the Cublington area, someone who had no legal business to be handling explosives had arranged for a quantity of gelignite to be brought to a farm on the edge of the Cublington area. A hundred and fifty pounds of the stuff, which had been obtained from a source in London, was delivered to the farm. It was to have been used to blow up culverts and road bridges in the vicinity of the airport as soon as any officials or contractors arrived in the area. Such actions would certainly have 'choked off' general sympathy for the Cublington cause. With the activities of extremists in Northern Ireland and of the so-called 'Angry Brigade' in England, this was a year in which political bombing was at its height 'only parallelled,' according to an official report, 'by that during the Fenian outrages in 1883/84 and that during the IRA outrages in 1938/39'.[2] In such circumstances the militant defenders of Cublington would have gained little sympathy.

Behind all these plans for desperate violence were two large assumptions. One was that the ordinary people needed to be encouraged or taught how to resist eviction. The other was that WARA's campaign of resistance would collapse once the Government and Parliament had decided that the airport should go to Cublington. Both assumptions were wide of the mark. The ordinary householders of Stewkley, Cublington, Dunton and Soulbury may well have abhorred the sort of violent resistance that the Home Defence Group had in mind, but few of them would willingly have moved. One has only to read the letters sent by them to the Roskill Commission to understand that their way of life—their attachment to homes, land, villages, friends and kin—was simply not transferable to another locality, however lavishly and carefully planned. They could not conceive of life elsewhere. They would have sat and awaited the bailiffs in a spirit of defiance or of uncomprehending fatalism. It is inconceivable that the people who had organised and supported the mass protests of WARA would have stood aside to let these evictions happen. If the situation at Cublington had ever come to it, there would have been organised, large-scale and completely passive resistance. And it would have been effective.

Many of the people in North Buckinghamshire thought that there had been no precedent for such resistance, but there had been and it was not so long ago. In the early 1930s the farmers, farm labourers and villagers of some parts of Suffolk and Norfolk resisted the tithe-gatherers by organised non-violence. Many of them admittedly lost their livestock and possessions in the process but they won a victory in the end. The Suffolk 'Tithe War' resulted in the setting up of a Royal Commission and the subsequent reform by Parliament of the Church of England's historic right to levy tithes in perpetuity from farmers. The

East Anglian campaigners did all this by completely legal and non-violent means. They disarmed their own hotheads and they rejected help offered them by Mosley's Blackshirts. The people of Cublington could have won their last-ditch battle by similar tactics. They had no need of their own hotheads and even less of 'Japanese-style' aid from outside. But of course it never came to that. The WARA campaign in Parliament and the press had contributed to a peaceful victory. At Cublington it was, from every aspect, a victory for good sense.

XIII
CAMPAIGNING FOR THE ENVIRONMENT

Problems of the environment, like the poor, are always with us. It is not only the prophets of global doom who have made us aware of them, with their apocalyptic vision of an overbred and underfed human race rushing like lemmings from barren continents filled with plastic and other technological detritus into the oil-and-detergent-covered oceans. The environment is near at hand and immediate. One cannot escape the word. It is, of course, a modish word, the bright new umbrella beneath which shelter many old concerns about the context of our daily lives. In Britain today, concern for the environment embraces by statute all road plans, sewage disposal schemes, building of all sorts, freight haulage, motor licensing, and any policy directives for the nationalised industries or sport, because all of these are responsibilities of the so-called Department of the Environment. The DOE was set up, however, partly in response to the same set of urgent problems that caused the upsurge of popular concern—the apparent need of our society to accommodate more households, more consumers, more cars, more leisure facilities, more air travellers, ever more of everything in the same restricted land mass.

'The result,' wrote Roy Gregory in his book, *The Price of Amenity*, 'is a well-known catalogue of developments, so often chanted nowadays as almost to sound hackneyed. Motorways and airports, power stations and overhead transmission lines, mineral workings and reservoirs, gasholders and natural gas terminals, New Towns and housing estates, factories and oil refineries, all invade the countryside and change the face of cities and towns. Necessary and desirable they may be; but, as often as not, these developments interfere with the amenities of the locality chosen for the project. Indeed their impact is frequently felt far beyond the particular patch or strip of land on which they are situated.'[1]

It is no longer possible to equate a technological threat to the environment on this scale with that old enemy of conservationists, 'the March of Progress'; in an inflationary world it is all too evident that we have to go on producing and earning, building and rebuilding, so that next year we may continue to enjoy this year's standard of living. That is one aspect of our modern concern for the environment. Another is that it represents an impatience with government, a desperate desire from below to participate in the control of the new forces which are changing the circumstances of day-to-day living. It is also a self-propagating concern. As one group of people on one side of the country protests against a proposed motorway or airport, their campaign is watched on television or through the newspapers by groups elsewhere, who are just as concerned about the local effects of other motorway and airport plans. We too have an environment, say the watchers: if they can fight to keep theirs, we can do the same for ours. There cannot now be many areas of Britain which are not jealously watched over by at least one amenity society.[2]

What did Cublington contribute to this wider environmental concern? Its supporters certainly saw it as a turning point in the national struggle. 'I would like to think,' said WARA's chairman, Desmond Fennell, 'that Cublington 1971 will go down as the year and place where ordinary people rose up in revolt against the tyranny of technology.' Many others saw it in similar terms and in the later stages they tended to take their cue from Professor Buchanan. He described the choice of putting the airport at Cublington or Foulness as a 'critical test' of the nation's sincerity in planning its commercial development with social purpose and in safeguarding its environment. To choose Foulness in preference to Cublington or any other inland site, he said, would 'show that this country at any rate, in spite of economic difficulties, is prepared to take a stand'. As we know, the country—that is to say the Government—did take a stand and this has encouraged others to take their stand in defence of other sectors of the environment. Inevitably, these people look to Cublington as an example of successful environmental concern. They seek to emulate its tactics in the hope of winning similar political deliverance from the decision of a council, a public inquiry, a nationalised industry or even a government department.

There is certainly much in the Cublington story that others can find of relevance to their own causes. First the organisation of WARA, its leadership structure, its flexibility, its ability to learn from its ordinary supporters, ultimately to hand over the public initiative to these supporters and yet still to control some of the more impatient ones. Then its mobilisation of local support: 62,000 individual members, the backing of two county councils, 125 parish councils and 85 other bodies, and the raising of £57,000 from individual donations and two years of tireless group activities. There was the expert presentation of the WARA case to the Roskill Commission, which used up most of these funds to enlist the services of the best lawyers and noise and planning consultants available. Not forgetting, of course, that the people of Cublington also became their own experts in matters of planning and aviation and that the campaign made full use of the talents of ordinary people. The professional man's skills were important, but so was usefulness with a hammer or a paint-brush or a baking tin. This became particularly evident when the Roskill decision went against them and the people of Cublington were forced to make their point not through barristers and experts, but to the nation at large through politicians and the press and television. Cublington's success in making its local anxieties into a matter of national, even of international, concern needs no elaboration here. Finally, there was the crucial political battle, fought by WARA in co-operation with a group of Members of Parliament whose case aginst an inland airport site it had nurtured and fed with information. The Cublington story abounds with examples which are of immediate and practical usefulness to other environmental pressure groups.[3] It also provided some broader lessons for the environment.

It would be a mistake, however, to try to construct from the Cublington experience an all-purpose, cookery-book recipe for successful environmental action. There always was a tendency to do this on the part of some journalists. Their basic attitude towards the campaign was not dissimilar from that of their colleagues who accused WARA of being 'almost disturbingly professional' in its methods or of purchasing its professionalism from a public relations firm. What was ignored by these critics was the feeling behind the display of skills, the people's attachment to the environment they fought so expertly to keep. To ignore the element of commitment is to make the story of any local resistance campaign into a travesty of what it is really about.

It would, in any event, be a sad day for a society if a minority group were able to prevent or delay the building of a public utility like an airport simply by a brilliant tactical resistance. In fact, it is highly unlikely that any such group could actually *prevent* construction of a major public work for, as with the Roskill Commission's terms of reference, the need for the project is seldom open to question. What the local pressure group can aim at is the selection of an alternative site for the development. To achieve this aim, it must be able to demonstrate that the social and environmental impact of the development will be less damaging if the alternative site is chosen and, hopefully, that the 'total cost' to the community will be smaller. The right of local objectors to suggest and argue for alternative sites is of prime importance, as the Stansted inquiry at Chelmsford showed, although it is often eroded by the lack of official information available to the objectors. The Cublington objectors, of course, did not have to produce their own alternatives since the Roskill Commission was considering evidence on four short-listed alternative sites. But initially, at least, WARA was handicapped by a serious lack of official information and it overcame this in ways that should prove interesting to other environmental groups.

The telling strength of the WARA case was that it was able to argue, not just negatively against an airport at Cublington, but positively for the selection of an alternative location at Foulness. In the final analysis, this counted for more than the campaign's other strengths—that it was led by men who became shrewd lobbyists and skilled at public relations, or even that it had strong backing from the local population and from many supporters farther afield. The key to WARA's success was its ability to demonstrate that the overwhelming weight of argument about sound planning policies, about the environment and about social justice came down in favour of Foulness. As one local newspaper put it succinctly: 'The best place for the third airport is Foulness. For in spite of the extra cost, this is the site where an airport will create least disturbance through noise, least disturbance of London's surrounding countryside and least disturbance of people's homes.'[4] That, in a nutshell, was the WARA case. It became the theme of Professor Buchanan's dissent from the Roskill majority, it was also the opinion of the 218 MPs who signed the Commons' anti-inland airport motion and, finally, it was a view which the Government accepted.

With the benefit of hindsight, it may appear that WARA's task was an easy one; it had only to follow the pro-Foulness line of argument unswervingly and all responsible opinion in the country would rally to its side. But it was not so straightforward. Throughout the Cublington campaign, WARA was brought face to face with an unwelcome feature of campaigning for the environment, that it can be a highly competitive and divisive activity. In general terms, this divisiveness can occur at many levels: between the local residents who oppose a development and those who welcome it, between the interests of different alternative sites and between those who feel threatened by a proposed development and those who stand to gain relief from it (like the noise sufferers at existing airports or the townspeople who cry out for a by-pass to divert the juggernaut lorries from their narrow High Street). The WARA leadership was not seriously worried by local support for the airport. For most of the campaign, they were able to avoid friction with other airport amenity groups and make common cause with the noise protesters at Heathrow and the resistance groups at the other Roskill sites. The testing time came during the Stage V public hearings, when the association found itself driven into a corner by the Commission's refusal to see the issue simply as a choice between a noisy airport at Cublington and a quiet one at Foulness. Mr Justice Roskill pressed WARA's counsel, Niall MacDermot, to say whether the association would rather that the third London airport were delayed than that it should be built at Cublington. 'Mr MacDermot, how far do you go?' he asked. 'Do you submit we ought to continue making Luton noise worse, Heathrow noise worse?' Mr MacDermot replied that the limits to be imposed on the traffic at existing airports was a policy decision for the Government. But the Commission remained unconvinced and, in the final Report, it incorporated the noise effects of expanded operations at Luton among the penalties to be suffered if Foulness were selected. From then onwards, WARA virtually lost the support of the Luton and District Association for the Control of Aircraft Noise (LADACAN). There was a potentially more damaging exchange later in the hearing when Mr Justice Roskill urged WARA's counsel to leave Foulness out of the reckoning and to choose between the inland sites. With some hesitation, WARA advanced arguments in favour of Thurleigh, the Bedfordshire site, thereby putting in jeopardy not only its relations with BARA but also its own support in the northern part of the Cublington area.

WARA was, of course, manoeuvred into this invidious position by the Roskill Commission and the damage to its relations with BARA was in time repaired in the formation of a common political front against an inland airport. The whole episode, however, goes some way towards explaining why WARA was accused by Anthony Crosland of giving the impression that it did not 'care tuppence for anybody else's noise and environment'. There is an issue of some importance to be examined here. Largely on account of the vigour and the scale of its resistance, WARA more than any other environmentalist group found itself accused of lacking a sense of social responsibility, of being middle-class in its values and of being run by top people. Mr Crosland himself brought all these criticisms together in a general critique of the conservationist lobby, which must in part have been aimed at WARA since it was published at the height of the Cublington debate:

'To say that we must attend meticulously to the environmental case does not mean that we must go to the other extreme and wholly neglect the economic case. Here we must beware of some of our friends. For parts of the conservationist lobby would do precisely this. Their approach is hostile to growth in principle and indifferent to the needs of ordinary people. It has a manifest class bias, and reflects a set of middle and upper class value judgements. Its champions are often kindly and dedicated people. But they are affluent and fundamentally, though of course not consciously, they want to kick the ladder down behind them. They are highly selective in their concern, being militant mainly about threats to rural peace and wildlife and well loved beauty spots; they are little concerned with the far more desperate problems of the urban environment in which 80 per cent of our fellow citizens live.'[5]

Mr Crosland's quarrel was in part with 'anti-growth' economists like J. K. Galbraith and Ezra Mishan: it is doubtful if many rank-and-file conservationists think in those academic terms. His remarks about the alleged class-bias of conservationist groups, however, reflect a commonly held view. They are similar to the comments of another Labour Party observer, James Curran, who ran the pro-airport campaign at Thurleigh and, after the Roskill recommendation, briefly took up the cause of an airport at Cublington:

'The basic assumption of the Roskill cost/benefit analysis is that there is a conflict of interest between the nation which needs an airport and the local community which can't stand the idea. Really the conflict is only between a minority and the national interest. Most of the objectors belong to the middle classes who believe that they have much to lose and little to gain.'[6]

To take the point about class composition first, conservation is obviously more likely to be a concern of those sections of the population that have achieved a degree of affluence than of those still engaged in a hand-to-mouth existence. To that extent, it is an activity of the 'haves' rather than the 'have-nots'. It is also obvious that some conservationist groups present to the world a solidly middle-to-upper class image. They appear to be run by people with time on their hands—by retired service officers, by housewives whose children have grown up, by members of a few landed families who, like characters in a Jane Austen novel, are able to spend their days in the leisured enjoyment of an inherited environment. That is, rightly or wrongly, the image that one associates with national organisations like the Council for the Protection of Rural England and with local organisations like the Friends of the Vale of Aylesbury. Indeed, the latter's list of vice-presidents (which, remarkably enough, included Douglas Jay) would do credit to any top people's social gathering. It does not necessarily follow that either the Friends or the CPRE can be accused of callousness towards the problems of 80 per cent of their fellow-citizens. Yet this degree of upper-middle-class predominance is not true of all environmentalist organisations. WARA and the NEW&EHPA at Stansted both embraced avowedly populist and 'progressive' groups within their organisations, as well as trying to enroll among their leaders the cream of local prestige and influence.

Yet accusations of class-bias are largely irrelevant, for in an economic and educational sense modern Britain is overwhelmingly middle-class. Certainly, the people that any community or group chooses as its leaders tend to be of a better 'class' than the rest—better educated perhaps, certainly more thrusting, more experienced and more articulate. This is a feature of all group activity. The English Peasants Revolt was led by a preacher and, among the leaders of the Russian Revolution of the proletariat, Lenin was a former aristocrat, Trotsky was a middle-class intellectual and Stalin a former theological student. Mr Crosland and Mr Curran (who lives in a Jacobean manor house) are clearly not working-class, but that is not sufficient reason to accuse the Labour Party as such of being middle-class. A more real division within the WARA membership had less to do with social class than with education. The campaign's leaders, the people who took the day-to-day initiative, who gave interviews to the press and television, were naturally enough the more articulate ones. The less articulate members of the Cublington community, however strongly they may have felt about the airport, seldom came into the limelight. If there is a criticism to be made against WARA, it is that the association did not give these less articulate supporters enough to do. It was not until the closing stages of the campaign with the series of WARA 'happenings' that the services and talents of these supporters were recognised and at all fully employed.

The criticism about middle-class *values* is more problematical for, in a sense, the environment is a wholly middle-class concern. If there was a dominant social type among WARA members, it was the lower middle-class family man—whose position on the social ladder mentioned by Mr Crosland was too recently attained for him to be thinking of kicking it down behind him. The distinction between this man and those trade unionists of Stansted and Thurleigh who welcomed the airport for the increased material prosperity, was not so much a matter of class or affluence, as of personal values. There was, in fact, a substantial number of trade unionists and industrial workers in the Cublington villages. Undoubtedly many of them chose to live where they did for the sake of their rural environment and they thought the inconvenience of longer journeys to work was a price worth paying. In any case, as Professor Buchanan pointed out at a CPRE meeting in Milton Keynes in May 1971, all rural areas tend to show a lack of jobs and facilities and to see a ready solution to this in the construction of a four-runway airport would mean that the third London airport could be built in almost any piece of countryside. He added that he did not consider it arrogant to think that middle-class values would be extended more generally in society—through rising incomes, increased house ownership, more leisure time and higher educational standards. All of these arguments, of course, leave out of consideration the obligation to posterity and the likelihood that, middle-class or not, the values of today's conservationists may well be something for which posterity will bless this generation. Posterity must be numbered among the 'have-nots' of the argument.

The other matter that arises from Mr Crosland's critique is whether one ought to expect any conservationist or other local pressure group to exhibit the general concern for society's problems which he expects. One must, of course, agree with him when he says, later in the same Fabian pamphlet, that 'since we have many less fortunate citizens, we cannot accept a view of the environment which is essentially élitist, protectionist and anti-growth'. Society as a whole cannot accept such a view, but it must be the task of Society itself and particularly of its political leaders to achieve a balance between the conflicting claims of industrial renewal and of conservation. It goes almost without saying that the many amenity groups that have sprung up in recent years do not exist to pursue these broader political questions. They came into being in response to specific and local threats to the environment. Often they were set up in haste, almost in panic, to do something before it was too late and the bulldozers arrived. To expect them to take a lofty, national view of their predicament is almost as Utopian as to expect trades unions always to bear in mind the national interest in whatever industrial action they take. It is not the primary purpose for which they were founded. A further problem is that very often the local conservationist cannot see a direct link between the motorway or airport which, in his opinion, needlessly threatens to demolish his village and the task of eradicating slum dwellings in another part of the country. The connection may exist, but it is seldom pointed out.

A similar point to that made by Mr Crosland was put to the Country Landowners' Association by Mr Edward Heath. Those who live in the country and love it dearly, he said, could not only help by handing it over to the next generation fresh and unspoilt; they could help also in supporting the building, for those less fortunate, of towns and cities where life could be as full and meaningful as in the country.[7] What saved Mr Heath's strictures from being equally Utopian was his statement that 'only the Government has the authority and the resources to change the environment for these people. Only government can replace the slums and build new schools and hospitals.'

At its base, campaigning for the environment is the product of a lack of confidence in government. It is because many people feel that their deeply-held values about the context of their daily lives are not only not shared by Authority, but not even comprehended, that there has been such an upsurge in recent years of protests, campaigns and threats to take on the authorities if necessary by direct action. At a crucial stage in the growth of this impatience at official policies towards the environment, when the Government was proposing to build the third London airport at Stansted, its competence and good faith became suspect in other ways. It was accused of high-handedness and shoddy planning. At the time it seemed to Labour Ministers that the only way out of this politically uncomfortable position was to hand over the planning of the airport to an impartial commission of inquiry, which would use the finest blend of judicial and economic techniques to arrive at a decision which everyone could respect. The result was the Roskill recommendation of Cublington, which nobody respected. Enough has been said elsewhere in this book to indicate the main criticisms that have been levelled at the Roskill Commission, at its terms of reference and methods, and particularly at its use of the cost/benefit analysis approach.

When it was appointed in 1968, the Commission was seen as the forerunner of the Planning Inquiry Commissions embodied in the 1968 Planning Act. It is doubtful, however, if anything like the Roskill Commission will be tried again within the foreseeable future. One cannot, for example, imagine seven men as eminently qualified as the Roskill Commissioners giving up three years of their lives to a similar undertaking, with the possibility that at the end of it all the Government will turn their judgment on its head.

The heavily judicial atmosphere in which the Roskill hearings were conducted affected WARA—and to a lesser degree the anti-airport groups at the other sites—in two main respects. First, the association was forced to decide whether it should bid for the best legal advocates it could find and thus present its case before the Commission on an equal footing with the local authorities and other public bodies. It decided that it should and it retained the services of two of the best planning advocates available. The desirability of employing barristers in public planning inquiries is always a matter of some debate, although there is no doubt that WARA's counsel more than justified the faith placed in them by their expert handling of the many complex arguments that arose in the Roskill hearings. Yet it was a formidably expensive matter for WARA, which spent £43,713 on its legal and technical representation. Indeed, the people of Cublington paid thrice over to keep the airport from their door—first as taxpayers (the public cost of the Roskill inquiry was £1,131,000), then as ratepayers (Bucks County Council spent £40,000 on its legal representation) and then as members of WARA. If they also happened to be members of the National Farmers' Union (which spent £6,000 on being represented before the Commission) they paid four times over. As the law in Britain stands at the present, there is no provision for legal aid at public planning inquiries and costs are awarded only in a relatively small number of successful objections to compulsory purchase and similar orders. The Roskill Commission was not, in any case, a statutory inquiry and so no costs could have been allowed. Taken together with the planning blight that had overshadowed the Roskill airport sites for more than two years, this seemed to many people to constitute a gross injustice. Certainly, if the Roskill type of inquiry is ever to be repeated, there ought to be provision for costs to be awarded to individual and group objectors.

The other local effect of the Commission's long-winded and heavily judicialised procedures was that it forced the amenity groups to put all their eggs into the one basket of legal argument and to refrain, for two years, from political campaigning. On a number of occasions, Mr Justice Roskill made it plain that the Commission took a dim view of any attempt to side-step the complex arguments about cost/benefit and to resort to any form of political search for a solution. Indeed, this was an almost inevitable consequence of the appointment of a judge as chairman of the Commission. The present Lord Chief Justice, Lord Widgery, has outlined the criteria which the judiciary should apply to inquiry commissions they are invited to lead. In a speech at the Mansion House, he said that the inquiry must be essentially judicial in nature, even if it has political overtones, but that the line should be drawn where the issue is political in nature and where the functions of the judiciary and the executive would be blurred.[8] Measured against these criteria, the decision to appoint a High Court judge to lead the third London airport inquiry can now be seen to have been a mistake. The inquiry involved a number of difficult areas of government policy which had not at the time been subjected to anything like full political debate. The Government's planning strategy for the South-East was still unknown, there was no national policy for airport location and the crucial question of whether an inland site for Britain's main twenty-first century airport would be politically acceptable remained unanswered. There were, of course, other questions of a political nature that arose during the arguments about cost/benefit analysis: for example, how much the nation should be prepared to spend in order to conserve things that have no apparent monetary price, like the countryside or ancient buildings or human communities. The Commission was operating in a political vacuum.

Professor Peter Self said that the Roskill inquiry 'may well go down in history as the one and only attempt to deliver a judgement of Solomon according to cost/benefit figures'. But King Solomon's judgement was swift; Mr Justice Roskill's took all of three years. For much of that time, the people whose ways of life hung in the balance were subjected to strain and even financial hardship. They knew instinctively that their fates could be settled only by a political decision and at Cublington especially they showed a strong desire to force this political decision into the open by means of demonstrations. The WARA leadership was reluctant to let this happen and for some time there was a gulf between some of the association's more frustrated supporters and the leadership which was prepared to go along with the extended legal argument about the *minutiae* of cost/benefit factors. This partly explains the great outpouring of popular feelings that took place at Cublington, once the Roskill inquiry was out of the way and the issue had become unequivocally a political one.

The essentially political nature of the airport campaign is the main lesson that Cublington can teach other campaigns for the environment. To some critics of WARA, it seemed that the association came very near to breaking the democratic rules of the game. But the game of environmental concern is one in which the rules are only gradually being formulated. The issue has not yet become political enough. The only means by which a democratic society can adequately choose between what it must destroy and what it should conserve, between the conflicting interests of one local community and another, between the contemporary economic interest and interests of posterity, is by political debate on the broadest possible base. If the debate is to succeed in drawing up the guidelines of a national policy for the environment, it must proceed not only at each individual planning inquiry and in the activities of amenity pressure groups, but also in Parliament, at the party political conferences, in the trades unions, professional organisations, academic seminars and, above all, in the communications media. The people of Cublington entered this debate very near to its beginning. Their primary concern was for the defence of their own environment and for their own survival as vigorous and viable communities. But it was not only in a spirit of self-interest that they made use of the press and television to put their local concerns before a larger audience; they were not 'manipulating' the media so much as co-operating with it in the telling of a story which in its wider aspects touched everybody's lives. Of course, the people of Cublington made mistakes and at times over-reached themselves in their desire for publicity. They began their campaign as novices and they won it before they became too professional. The same was true of their conduct of the political campaign. They began as people with very little understanding of how Parliament works or of how to lobby its Members. Again they learned from their own mistakes and from the desire of MPs to learn from them. The people of Cublington played an important pioneering role in making the environment into a major political issue. The great merit of their campaign was their ability to see the larger issues that it raised for the nation, without at the same time forgetting or minimising the plight of their ordinary, very anxious supporters.

NOTES
and
INDEX

NOTES

CHAPTER I

1. March 24 1971.

2. *Permission to Land* by Brian Cashinella and Keith Thompson (Arlington Books 1971), p. 14.

3. *Roskill Report*, p. 154, para 28.

4. *The Airport and Airways Development and Revenue Act*, 1970.

5. John Wilkinson (Con. Bradford W.), *Hansard*, July 28 1972.

CHAPTER II

1. For a full explanation, see 'Aircraft Noise' by Geoffrey Holmes, Public Health Inspector of New Windsor, reprinted from *Environmental Health*, November 1971.

2. *Noise: Final Report*, Cmnd 2056 (HMSO 1963).

3. *The Observer*, March 9 1969—after at least two changes, the telephone number for noise complaints is now 01–836–1207 ext. 1107, but no doubt it will change again.

4. *Roskill Report*, p. 57, para 7.2.

5. *Hansard*, August 9 1972.

6. *Roskill Report*, p. 57.

7. *History of the Second World War: Works and Buildings* by C. M. Kohan, HMSO 1952, pp. 279–88.

8. *A National Airports Plan* by Rigas Doganis (Fabian Tract no. 377, November 1967).

9. *First report from the Select Committee on the Nationalised Industries 1970–71*: the British Airports Authority (House of Commons Paper no. 275, HMSO), see p. 37, Question 151, and p. 42, Question 167.

10. *Hansard*, July 28 1972.

CHAPTER III

1. *Roskill Report*, p. 1.

2. *Fifth Report of the Estimates Committee* (Session 1960–61), House of Commons Papers 233, HMSO.

3. *Report of the Interdepartmental Committee on the Third London Airport*, CAP 199, HMSO 1964.

4. Cook, *The Stansted Affair* (Pan 1967), p. 19.

5. Cook, *op. cit.*, p. 25.
6. *Town and Country Planning Act* 1962, s. 199.
7. *Public Inquiries as an Instrument of Government* by R. E. Wraith and G. B. Lamb (Allen and Unwin 1971), p. 206.
8. *Hansard*, June 29 1967.
9. Patrick Gordon Walker, *The Cabinet* (Jonathan Cape 1970), p. 168.

CHAPTER IV

1. *The Cabinet*, p. 169.
2. *Roskill Report*, p. 11, para 3.6.
3. *Roskill Report*, p. 12, para 3.10.
4. *Hansard*, May 20 1968—strangely, the terms of reference are not quoted anywhere in the Roskill Report.
5. *Hansard*, May 20 1968.
6. *Roskill Report*, p. 11, para 3.6.
7. *Roskill Report*, p. 12, para 3.11.
8. *Roskill Report*, p. 22, para 4.34.
9. See 'The Generation and Coarse Evaluation of Alternatives in Regional Planning' by Ted Kitchen, *Journal of the Royal Town Planning Inst.*, vol. 58, no. 1, (January 1972), pp. 8–12.
10. *Roskill Report*, p. 22, para 4.35.

CHAPTER V

1. 'Disruption of Community Life', by Dr P. M. Abell, C. Bell and P. D. Doreian, *Roskill P. & P.*, vol. VIII, pt 2, s. 4.
2. *Disruption Study*, p. 9.
3. *Roskill Report*, p. 152, para 17.
4. *Chronicles of Whitchurch, Bucks* by G. W. Wilson (privately printed 1909).
5. *Stewkley, Bucks, a brief history* by Kate Mayne and W. G. Capp (1955).
6. Haddenham Parish Council evidence, *Roskill P. & P.*, vol. V, p. 1128.
7. *Pipe Rolls, 1189–1199* (County Museum, Aylesbury 1923), *Eyre Roll 1227* (Bucks Archaeological Society 1945), *Feet of Fines* (BAS 1940).
8. *Subsidy Roll 1524* (Bucks Records Society 1950).
9. *Methodist Recorder*, January 21 1971.
10. *Roskill P. & P.*, vol. V, p. 1223.

CHAPTER VI

1. See Arthur Percival, *The Organisation of an Amenity Society* (Civic Trust 1967), p. 22.
2. *Roskill P. & P.*, vol. VIII, pt 2, s. 4.

CHAPTER VII

1. *Roskill Report*, Appx 4, p. 170.
2. See *Roskill P. & P.*, vol. III.

3. *Roskill P. & P.*, vol. IV.
4. *Roskill P. & P.*, vol. VI.
5. *Roskill P. & P.*, vol. V.
6. *Roskill Report*, Appx 4, p. 177, para 58.

CHAPTER VIII

1. See Chapter V above.
2. *Roskill P. & P.*, vol. VIII, pt 2, s. 4.
3. *Roskill P. & P.*, vol. VII, pt 2.
4. *Forestry in Great Britain*: an Interdepartmental Cost/Benefit Study (HMSO 1972).
5. *Roskill Report*, p. 12, para 3.11.
6. *New Society*, January 28 1971.
7. *Roskill P. & P.*, vol. VII, pt 1, p. 6.
8. Peter Self, 'Cost-benefit Analysis and the Roskill Commission', *Political Quarterly*, vol. 41, no. 3 (1970).
9. Martin S. Feldstein, 'Cost-benefit Analysis and the Public Sector', *Public Administration*, XLII (1964).
10. *Roskill P. & P.*, vol. VII, pt 1, p. 12.
11. *The Times*, April 2 1970.
12. Peter Self, 'Cost-benefit Analysis and the Roskill Commission'.

CHAPTER IX

1. *Roskill Report*, p. 130.
2. At a planning conference April 27 1972, quoted in Tony Aldous, *Battle for the Environment* (Fontana 1972), p. 250.
3. *Hansard*, May 20 1968.
4. *Roskill Report*, p. 137.
5. *Hansard*, April 26 1971.
6. *The Times*, January 22 1971.
7. *Roskill Report*, p. 52.
8. *Roskill Report*, p. 55.
9. *New Society*, January 28 1971.
10. *A Criticism of the Final Report of the Roskill Commission* (WARA), February 10 1971, p. 13.
11. *Roskill Report*, p. 155.
12. *The Times*, January 22 1971.
13. *Roskill Report*, p. 159.

CHAPTER X

1. Quoted in *Campaign*, April 30 1971, p. 34.
2. *Ibid*, p. 34.
3. *New Statesman*, December 25 1970.
4. See 'The Class-conscious Search for a Third Airport', *Sunday Times*, April 5 1970.

CHAPTER XI

1. *Hansard*, April 26 1971.
2. *The Times*, October 28 1971.
3. *Hansard*, August 7 1972.
4. *Hansard* (House of Lords), February 23 1971.

CHAPTER XII

1. See Chapter VII.
2. *Report of Her Majesty's Inspectors of Explosives for 1971* (HMSO), p. 34.

CHAPTER XIII

1. Roy Gregory, *The Price of Amenity* (Macmillan 1971), p. 2.

2. By 1968 there were over 600 amenity groups in Britain, many of them of recent vintage, according to John Barr, 'The Amenity Protesters' (*New Society*, August 1 1968).

3. A comparison of WARA's methods with those of other environmental campaigns is being prepared by R. H. Kimber and J. J. Richardson of Keele University and is due to be published in late 1973. For basic practical advice, the best work is still Arthur Percival, *The Organisation of an Amenity Society* (Civic Trust 1967); for more devious advice see Anthony Jay, *The Householder's Guide to Community Defence against Bureaucratic Aggression* (Cape 1972).

4. *Oxford Mail*, January 22 1971.

5. Anthony Crosland, *A Social Democratic Britain* (Fabian Tract 404, January 1971).

6. Quoted in the *Sunday Times*, April 5 1970.

7. *The Times*, October 28 1971.

8. *The Times*, July 5 1972.

INDEX

INDEX